High Frequency and Microwave Circuit Design

High Frequency and Microwave Circuit Design

Charles G. Nelson, Ph.D.

Department of Electrical
and Electronics Engineering
California State University
Sacramento, California

CRC Press

Boca Raton   London   New York   Washington, D.C.

Library of Congress Cataloging-in-Publication Data

Nelson, Greg (Charles G.)
    High frequency and microwave circuit design / Charles G. Nelson.
       p.   cm.
    ISBN 0-8493-0249-8 (alk. paper)
    1. Microwave circuits--Design and construction. 2. Electronic circuit design.
    3. Modulation (Electronics) I. Title.
    TK7876.N46   1999
    621.381'32--dc21   99-40461
                                                          CIP

No claim to original U.S. Government works
International Standard Book Number 0-8493-0249-8
Library of Congress Card Number 99-40461
Printed in the United States of America 1 2 3 4 5 6 7 8 9 0
Printed on acid-free paper

Preface

"All right, Dr. Nelson," the potential reader may be thinking or saying, "what unfulfilled need does this book aim to fill?" Well, I say, every year we expect B.S. graduates in electronics engineering to have learned something that we formerly saved for graduate work. The material in this book is like that. The curriculum at this non-Ph.D.-granting institution requires a year of active circuit design from a book such as *Microelectronic Circuits* by Sedra and Smith. Thus, the student's exposure to high frequency effects in electronics will stop with the hybrid-pi high frequency equivalent circuit, leaving him or her to be most astonished and feeling helpless when they encounter a spec sheet which states scattering coefficients. Also, even though the curriculum might require an electromagnetics course treating wave behavior, the topics in electromagnetics books, even though they may treat standing waves and impedance matching, often do not mention that one's knowledge of the impedance to be matched may actually be stated in terms of scattering coefficients. Thus, a concise introduction to scattering coefficients that is understandable to undergraduates is a key aim of this book.

If there is one somewhat specialized audience at which this book is aimed, it would be those engineers who are or will be working to provide the world with the vast number and types of communication apparatus which will be needed at the beginning of the twenty-first century. These devices will have various digital functions and features, yet there will also need to be in them various aspects of analog design. The engineering graduate who thinks he or she will be able to confine herself, or himself, to either analog or digital design will quickly find that one has voluntarily limited one's usefulness to an employer. The author hopes to "demystify" some of the useful techniques that future communication engineers may need. He hopes that he can emphasize that many concepts have not changed and that the newer concepts and perspectives easily grew out of the traditional. One of his major aims has been not to startle the reader with many totally new or unfamiliar techniques.

Resonant circuits are a topic to which the student may have been introduced in both network analysis and active circuits courses. Yet, those presentations may both have been rather academic and totally lacking in relation to practical needs. The author hopes to motivate study of this topic by examples and exercises, which are applied to transmitters, receivers, and other applications requiring selectivity in the signals which are passed or rejected.

Chapter 1 begins at medium frequencies and quickly arrives at what are generally considered microwaves. It calculates the easily obtained distributed parameters in coaxial lines and determines their propagation constants and low and high frequency characteristic impedance. It moves on to the design of microstrip lines for specified characteristic impedance. Reflection coefficients and their effect on input impedances of lines are next derived, as well as the effects of attenuation on input impedance. Use of the Smith chart is reviewed. Finally, we arrive at scattering coefficients as two-port parameters for transistors. We are introduced to stability considerations and conditions for simultaneous conjugate impedance matching at input and output.

In Chapter 2, we begin the task of impedance matching by considering using L-sections of lumped impedance. We move on to quarter-wavelength and near-quarter-wavelength transmission lines for impedance matching; how and when we may use a matching section, which is not exactly a quarter-wavelength, is a straightforward procedure that is not commonly taught. Then we consider how one shunt or series reactance may match the transmission line if it is placed at the correct location. Finally, we see how the required reactance may be provided by an appropriate length of short or open-circuited transmission line, called a "stub." For completeness, we have also included irises for the matching of waveguides.

In Chapter 3, we begin with bandpass amplifiers using LRC circuit design and crystal filters. We apply these concepts to the needs of receivers and calculate performance limitations. We look at what are the requirements for a successful oscillator and illustrate it with perhaps conceptually the simplest oscillator, the phase shift oscillator. We move on to the increasing complexity of the Colpitts, the Hartley, and the crystal-controlled variety. Finally, we consider how one might design a voltage-controlled oscillator, as may be found in phase-locked loops.

Chapter 4 looks first at the most elementary ways of producing and detecting amplitude and frequency modulation. Synchronous detectors are the first demodulators considered, then later, the less expensive envelope detector. Finally, we look at the basics of the most common forms of digital modulation and demodulation.

Noise can be a very complex and theoretical topic. The approach here in Chapter 5 is simply to state the results using as given constants the arcane symbols of statistical mechanics. We define noise factor and noise figure in functional ways and look at the general way in which noise is added as we cascade amplifiers. Having established that overall noise performance depends mainly upon both the gain and the noise factor of the first stage, we learn how to draw amplifier noise and gain circles so as to have a systematic way to make such tradeoffs. We finish by seeing the high noise performance of FM and the markedly higher performance of systems of digital modulation.

Chapter 6 aims to lift, at least a little, the veil of mystery which has for many years hidden from young engineers some rather straightforward characteristics of antennas. The principles are introduced in such a way that a minimum amount of theory leads to a number of important performance characteristics. An aspect of this chapter as a capstone for the book is to use antenna gains, transmitter power, distance, carrier frequency, and bit rate in the evaluation of the optimum communication systems.

The computer revolution has also made available very powerful calculators which can do complex algebra, for reasonable prices. Appendix A gives a few helpful formulas which enable much of the impedance match problem to be soluble directly on the calculator.

Einstein was quoted that we should make complex things as "simple as possible, but no simpler." It is fondly hoped that this book can make the areas of high frequency design as clear as possible to the graduate electrical engineer.

About the Author

Charles G. Nelson, Ph.D., was born in Northport, Michigan. He received all his primary education in a tiny school district and graduated as co-valedictorian of his class. All of his academic degrees are in electrical engineering, including the Bachelor of Science from Michigan State University and the Master of Science and Doctor of Philosophy degrees from Stanford University. His doctoral research involved conversion efficiency of a klystron using varying profiles of magnetic field to hold the electron beam together. A summary was published in the *Transactions on Electron Devices of IEEE*. Other publications were in the Annual Convention on Engineering in Medicine and Biology and periodic and final reports to the California Department of Transportation (CALTRANS). This is his first textbook to be published, although his students have for years used his photocopied notes as the only textbook for two or three subjects.

His military service was with the Research Lab of the Ordnance Missile Labs at Redstone Arsenal, Huntsville, AL, where his primary assignment was the study of a communication system using pulse position modulation with only discrete positions allowed, thus exhibiting some of the advantages and limitations of digitization. He had industrial experience with Zenith Radio Research Corporation of Menlo Park, CA, working on extending the operation frequency of the electron beam parametric amplifier, and summer research with NASA at Ames Research Labs, doing early studies of phase modulation of a light beam in lithium niobate. His research interests continue to be in fields and waves and communication systems, and in sound and noise pollution.

Dr. Nelson has taught electrical engineering at California State University, Sacramento since February 1965. He served as chair of the department of electrical and electronics engineering from 1965 to 1967 and from 1979 to 1986. He has been active in various areas of faculty governance for many years. He has been registered by the State of California as a Safety Engineer. He is a member of Tau Beta Pi, Pi Mu Epsilon, and Sigma Xi, and a life member of the IEEE.

Dr. Nelson has been married to Nina Volkert-Nelson since 1967. They have two adult children, a son educated as a computer scientist and a daughter who has her Master's degree in cello performance. In his leisure, Dr. Nelson enjoys reading both fiction and nonfiction, from James Thurber to John LeCarre to Robert Woodward. He loves listening to a variety of instrumental and vocal music, is considered to be an opera buff, and he sings at the level of the better church choruses. He also enjoys the Pacific surf and wildlife and the tranquility of the shores of Lake Michigan.

Contents

1

From Lumped to Distributed Parameters

1.1 Expected and Unexpected Results from the Network Analyzer

One very useful instrument for studying high frequency behavior of circuits is called the vector network analyzer; an example of one is pictured in Figure 1.1.

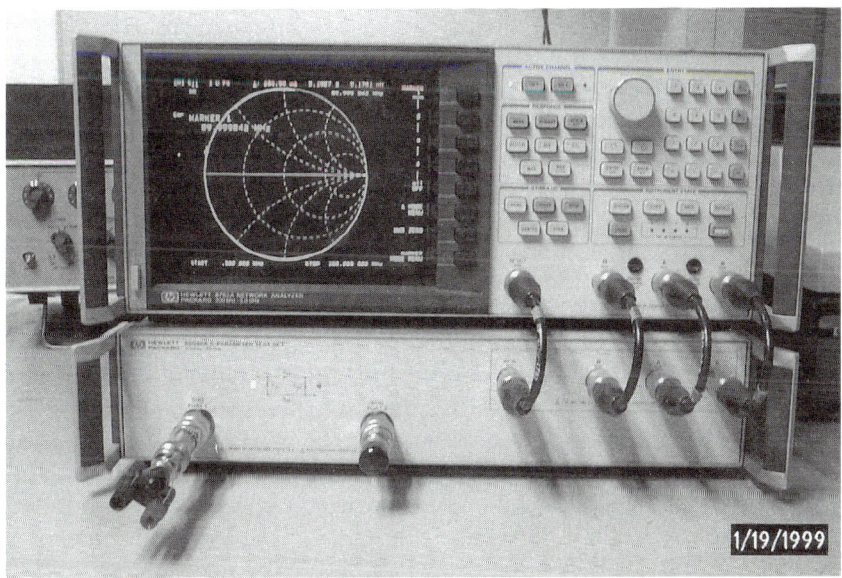

FIGURE 1.1
Vector network analyzer with adapters, capacitance load.

Measurements can be made of impedance connected to either port 1 or port 2, or of the amount of signal gain and phase shift from one port to the other, when an amplifying or attenuating two-port is connected between the ports. One first selects a frequency range of interest in the range from 0.3 to 3000 MHz, and then calibrates the instrument for that range and for the type of measurement planned. As an example of what is possible, the machine was calibrated from 0.3 to 100 MHz for measurements of the impedance connected to port 1. After calibration, a type N to BNC coaxial adapter was connected to port 1, and a BNC to binding post adapter was connected to the first adapter, after which a nominal 47 pF capacitor was connected to the binding posts without shortening the original capacitor leads.

The reactances measured were as follows:

Frequency	Reactance	Eq. Circuit Element
1.0 MHz	–j3050	52.2 pF
2.0	–j1535	51.8 pF
5.0	–j606	52.5 pF
10	–j302	52.7 pF
20	–j144.6	55.05 pF
50	–j38.18	83.3 pF
100 MHz	+j21.6	34.4 nH

Now, these data are not too startling at the lower frequencies, although one observes a modest bit of jumping around. The next thing done was to disconnect the capacitor and make measurements of the impedance of just the adapters that were added after calibration. Again, there was some jumping around by the data, but we could say the adapters provided a fairly consistent 6.0 pF. Hence, since the capacitance of the adapters and the capacitor being measured are effectively in parallel, they add, and we can probably say with good accuracy that the "official" capacitor provides very close to 46 pF, which is certainly well within 10% of the nominal value. But now we ask, "what on earth is going on at high frequency?" and we must suspect that the long leads are having an effect. Indeed, the total negative reactance is going down at high frequency faster than it would for constant capacitance, until eventually the reactance goes positive, i.e., inductive. So we must conclude that the long leads provide some inductance, the reactance of which cancels some of the capacitive reactance, giving a lower net reactance. Every little segment of wire contributes some inductance, so we say inductance is "distributed" uniformly along the leads.* The "smart" but naïve machine calculates a single equivalent circuit element which is a capacitor higher than the nominal value, until one reaches the frequency where the inductive reactance in the leads takes over and makes the total impedance inductive.

Example 1.1.1

The author once heard and has for years been quoting a rule of thumb that one can expect component leads to inject a nanohenry of inductance for each

---

* If the reader suspects that each little segment of wire also contributes a little bit of resistance, he/she would be right. It happens that the amount is so small as to be unobservable here.

millimeter of wire. Let's play with our data a bit and see how well this rule checks out. The reactances obtained above were each the average of 16 individual measurements, so we will assume them accurate and calculate the corresponding admittance (which is called "susceptance"). We will assume that the adapters add 6.0 pF and subtract that susceptance from the total. We convert the new susceptance back to impedance and subtract out the reactance corresponding to 46.0 pF. We will attribute the difference to inductive reactance and calculate the lead inductance from that. We will do this at 20 MHz and leave the numbers at 50 and 100 MHz for the exercises.

SOLUTION $-1/j144.6 = j6.93 \times 10^{-3}$
Now, 6.0 pF yields $Y = j7.54 \times 10^{-4}$ at 20 MHz.
   Subtracting gives $Y = j6.16 \times 10^{-3}$, or an impedance of $Z = -j162$.
   At 20 MHz, 46.0 pF yields impedance of $-j173$
   Therefore, a subtraction gives the result that the inductance is adding $j11$ ohms, which at 20 MHz yields an inductance of 87.5 nH. Because the lead lengths totaled about 55 mm, our result says to expect about 1.5 nH/mm of lead length.

EXERCISES 1.1
Repeat the calculations above for 50 and 100 MHz.
What inductance do you get?

ANSWERS   You will get 89.3 and 86.7 nH, respectively.
   The results above represent a departure from the type of circuit modeling the student of electrical engineering or electronics technology first sees, in which the circuit parameters are assumed, with good justification, to be "lumped" in a fairly obvious location. The capacitor in the example above appeared physically as a lump of dielectric material with leads sticking out of two ends. One is normally taught that these leads are basically short circuits, not contributing any resistance or other impedance. What we found out was that the leads contributed small amounts of inductance, but that at high frequencies the impedance of the inductance became higher than the impedance of the intended capacitance. Sometimes the frequency at which the reactances are equal is called "frequency of self-resonance."

1.1.1   Interwire Capacitance in an Inductor

Once one begins thinking about accidental circuit parameters which may appear, one might wonder, "suppose we set out with wire and a coil form to wind an inductor. Can we expect every bit of wire to contribute capacitance to every other bit of wire?" The answer is, "Yes, indeed." We will in fact provide a "recipe" for the rf engineer to fabricate handmade inductors because this is one component which may commonly be needed to be connected to two terminals of an integrated circuit in which many functions have already been provided. The procedure is discussed in *Reference Data for Radio Engineers,*

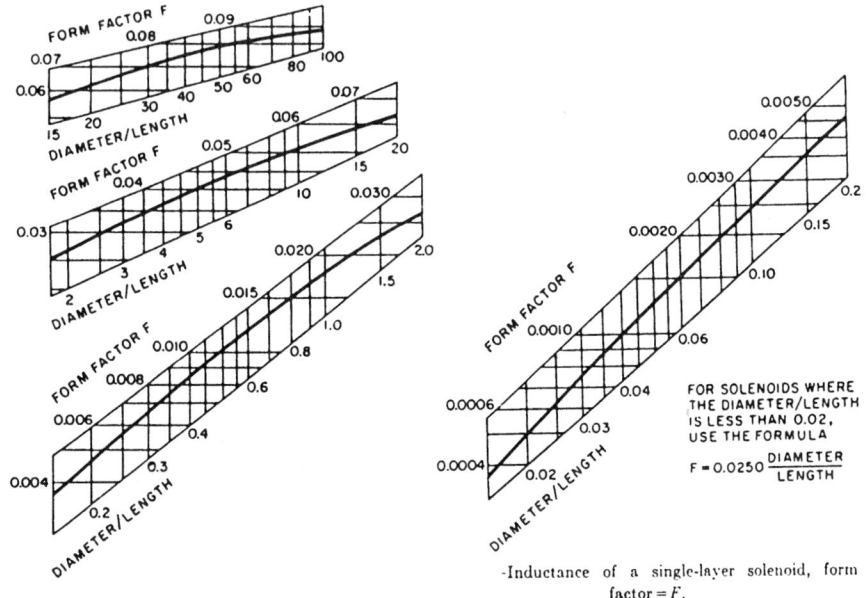

-Inductance of a single-layer solenoid, form factor = F.

FIGURE 1.2
Form factor for single-layer solenoids.

originally published in 1943 by the Federal Telephone & Telegraph. We illustrate several aspects of such design in the following example.

Example 1.1.2 Winding a coil for specified inductance

The author was hoping to wind a coil with inductance close to 0.5 µH.

He took AWG #12 wire because he had some and wanted to keep the resistance as low as possible, hence Q high. Also, the wire was stiff enough to avoid any accidental short circuits between adjacent turns of the wire. The temporary coil form was a ballpoint pen. Ten turns were wound. The resulting coil was 3 cm long with a diameter of 1 cm. Now, the handbook mentioned above predicts for a solenoidal coil, an inductance given by

$$L = n^2 dF \ \mu Henries,$$

where n is the number of turns, d is the coil diameter in inches, and F is a "form factor," depending upon the ratio of coil diameter to length, the form factor being given by the graph in Figure 1.2. The author read F = 0.0075 for diameter to length ratio of 3, d = 0.394 inch, so for n = 10, the inductance should be 0.295 µH. Network analyzer measurements at the lower frequencies were consistently 0.300 to 0.305 µH, so the procedure seems to work most satisfactorily.

Now, the next thing one might ask is, "what is the frequency of self-resonance and what capacitance does that represent?" From the network analyzer, self-resonance occurs at 119.3 MHz. Capacitance to be accounted for is

$$C = 10^6 / \left(2\pi \times 119.3 \times 10^6\right)^2 \times 0.3 \approx 6 \text{ pF}.$$

As can be seen from the previous example, the total capacitance can be accounted for as capacitance of the adapters connected to the network analyzers. Thus, we have not really found the wire-to-wire capacitance. Some indication of what to expect may be obtained from Krauss, Bostian, and Raab; in their book, *Solid State Radio Engineering,* they give graphs of the expected self-resonant frequency for RF chokes (a somewhat generic name for inductors meant to have high impedance at high frequencies). For a 0.3 μH choke, they predict resonance between 200 and 450 MHz. This would predict winding capacitance between 2.1 and 0.42 pF. The coil was not tightly wound, so expected capacitance might be nearer the lower value, and it is not greatly surprising that it apparently had no effect on the frequency of self-resonance.

EXERCISES 1.2
Predict the number of turns required for inductances of 0.5 μH and 1.0 μH, continu-ing with 1.0 cm coils. However, estimate a new number of turns, assume the coil length increases proportionately to number of turns, and converge to a number.

ANSWERS For 14 turns, the author finds F = 0.0065, L = 0.501 μH.
Trying to extend this to 1 μH, the coil gets rather long.
A second layer of wire is indicated, but the process above is supposedly only good for a one-layer coil.

1.2 Distributed Parameters in Coaxial Lines

We have seen how distributed parameters can arise in rather simple circuits. We next will see how they can become important in conductors that, at low frequencies, are expected to be perfect conductors for conducting generators to loads without intervening impedance. We first compute the distributed parameters in coaxial lines; coaxial lines were chosen because the symmetry of their geometry permits accurate results without a lot of finagling and further simplifying assumptions. Later, we will make some generalizations relative to other geometries. Consider the cross-sectional view of a coaxial line in Figure 1.3.

To obtain capacitance, we need to relate the voltage difference between conductors to the charge on the conductors. We assume the line to be effectively

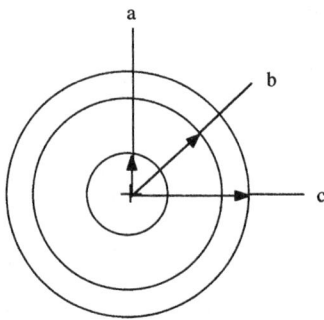

FIGURE 1.3
Geometry of a coaxial line.

infinite in length, so that our calculation is true anywhere on the line. It is not now feasible to talk about the total charge on one of the conductors, as it would be also infinite; instead, we assume that whatever charge is on a conductor is distributed uniformly along the length of the conductor with a density ρ_L Coulombs per meter. We apply Gauss' law over a cylinder of 1 meter length and radius "r," that is between "a," the radius of the inner conductor and "b," the inside radius of the outer conductor. Because the Gaussian surface is also coaxial with the conductors, the symmetry permits us to assume that the displacement D will be uniform over the Gaussian surface and will not have a z-component. Therefore, our closed surface integral has zero contribution from the ends of the cylinder and with a uniform D, the integral of D over the closed surface is simply the product of it and the area of the curved surface. We equate this to the charge enclosed by a meter-long cylinder and have

$$\oint D_r \bullet ds = D_r \, 2\pi r(1) = \rho_L(1); \text{ hence } D_r = \frac{\rho_L}{2\pi r}.$$

To get voltage between the conductors, we need to integrate $E_r = D_r/\varepsilon$; so,

$$V = \int_a^b \frac{\rho_L dr}{2\pi \varepsilon r} = \frac{\rho_L}{2\pi \varepsilon} \ln \frac{b}{a}.$$

Now, because normal capacitance is the ratio of charge to voltage, here we find the ratio of charge *per meter* to voltage and obtain farads *per meter*. We will find that **all** of the distributed parameters we work with will similarly be Henries *per meter*, ohms per meter, and so on. We will designate the capacitance per meter as

$$C_L = \frac{\rho_L}{V} = \frac{2\pi \varepsilon}{\ln \frac{b}{a}}.$$

The reader should remember that the quantity "c" is the product of a "rel ative dielectric constant" ε_r for the dielectric between conductors, times the so-called permittivity of vacuum ε_0. Because it is the only dimension in the capacitance expression which does not cancel out, clearly the dimensions of ε_0 may be considered to be farads per meter.

We next say, "electromagnetics theory tells us that if there is a magnetic field intensity 'H' resulting from currents in the conductors of a coaxial line, we can state a density of magnetic stored energy $\dfrac{\mu H^2}{2}$ Joules per cubic meter."

Of course, the circuit element which stores magnetic energy is called an inductor, so we next seek to find the total energy stored between conductors of a coaxial line. The simplest theory tells us that if there is a current in the center conductor, then an equal and opposite current flows in the outer conductor. We obtain H through the use of Ampere's law, which says that the line integral of H around a closed path gives the total current enclosed. Since a path which is totally outside both conductors encloses *no net current*, we can infer that H = 0 outside both conductors; there is therefore shielding of the "outside world" from currents flowing in the conductors, as long as they are equal and opposite. We should emphasize that certain naïve mistakes, such as connecting an antenna of balanced geometry to the basically unbalanced geometry of a coaxial line without using a "balun," a kind of transformer, can destroy the condition of equal and opposite currents and hence ruin the shielding effect.

Next, clever use of Ampere's law is to apply it on a path which is also coaxial with the coaxial conductors, along which symmetry assures that the value of H is a constant, and the line integral is simply the product of integrand and path length. So, when the path of integration has radius "r" which is greater than a and less than b, consider,

$$\oint H \bullet dl = 2\pi r H_\phi = I; \text{ therefore, } H_\phi = \frac{I}{2\pi r},$$

and the total energy stored between the conductors of a 1 meter length of coaxial line is the integral of energy density over the volume of the dielectric, as

$$W_m = \int_0^1 dz \int_0^{2\pi} d\phi \int_a^b \frac{\mu}{2}\left(\frac{I}{2\pi r}\right)^2 r dr = \pi\mu\left(\frac{I}{2\pi}\right)^2 \ln\frac{b}{a} = \frac{\mu I^2}{4\pi}\ln\frac{b}{a}.$$

We know of course that the energy stored in an inductor is $(1/2)LI^2$, so equating this to the expression above, we get the inductance expression

$$L \quad \frac{\mu}{2\pi} \ln \frac{b}{a} \quad \text{(in Henries/meter, as long as } \mu \text{ is in S.I. units).}$$

This is sometimes called "external inductance," because it is related to energy stored in neither conductor, but between them. There is of course non-zero magnetic intensity inside both conductors at low frequencies, but the so-called "skin effect" reduces such fields to insignificance at high frequency, and external inductance is the only significant inductance at high frequencies.

From an electromagnetics course, the reader may remember the skin effect. Basically, it is a phenomenon wherein at high frequencies, the current in a conductor crowds towards the surface of the conductor; if the conductor has an inside and an outside surface, as does the outer conductor of a coaxial line, the current flows in a thin layer along the *inner* surface, i.e., near the magnetic fields in the dielectric that electromagnetics scientists and engineers like to think of as the root cause of the currents. It turns out that one can accurately calculate high frequency resistance by assuming that the current is uniform in a thin layer called the skin depth and zero elsewhere. Thus, high frequency resistance of a coaxial line is given by

$$R_L = \frac{1}{2\pi\sigma\delta}\left(\frac{1}{a}+\frac{1}{b}\right) \text{ohms/m} .$$

Here we have implicitly assumed that the conductors are made of the same material and that σ is the material's conductivity in Siemens/meter, δ is the skin depth in this material at this frequency and is given by

$$\delta = \frac{1}{\sqrt{\pi f \mu \sigma}} \text{ m},$$

as long as all the quantities in the equation are in S.I. units. We should note here that permeability $\mu = \mu_0\mu_r$ where $\mu_0 = 4\pi \times 10^{-7}$ Henries/meter is the permeability of vacuum and μ_r is the relative permeability, which for some ferromagnetic conductors can be a number which is orders of magnitude greater than 1. The consequence is that strongly ferromagnetic materials may well have smaller skin depths than the best conductors, which are not ferromagnetic.

We mentioned before that there is small penetration of currents into the conductor at high frequencies. Small currents lead to small amounts of magnetic energy, hence the resulting *internal* inductance is small. At high frequencies, it may be calculated simply as $L_i = R/\omega$, and is normally very small compared to the *external* inductance which was derived above, but normal good practice might be to calculate it and justify neglecting it.

One more standard distributed circuit element is called conductance and results from the circumstance that standard dielectrics may have very small amounts of conductivity and other ways in which they may dissipate energy. Circuitwise, the best way to write this is as though the dielectric constant has a negative imaginary part, in addition to its usually known real part. Thus, we may write accordingly

$$\varepsilon = \varepsilon_u \varepsilon_i (1 - jd.f.),$$

Here, the normal quoted permittivity of the dielectric is the quantities outside the parentheses and d.f. stands for dissipation factor, which may be quoted for several values of frequency in handbooks which give the properties of commonly used dielectrics. Therefore, the convenient way to compute a transmission line conductance is

$$G_L = \omega C_L \times (d\,f),$$

Example 1.2.1

As we have given a number of formulas without applying any of them, now is a good time to put them to use. Let us calculate all the distributed circuit parameters at a frequency of 1 MHz for RG 58U coaxial line. "Official" numbers are inner conductor diameter $2a = 0.030$ inches, dielectric diameter; hence, the inner diameter of the outer conductor is $2b = 0.105$ inches, the dielectric is "stabilized polyethylene," for which the official dielectric constant is $\varepsilon_r = 2.26$, and the dissipation factor may be assumed to be 0.0002. Note that inches are not S.I. units, and must be converted to meters. Assume both conductors are copper, for which the conductivity is

$$\sigma = 5.8 \times 10^7 \text{ S/m.}$$

SOLUTION   Distributed capacitance is independent of frequency, and has the natural log of the ratio of dimensions, so we need not convert dimensions just yet.

$$C_L = \frac{2\pi \times 2.26 \times 8.854 \times 10^{-12}}{\ln(105/30)} = 100.4 \text{ pF/m}$$

Distributed external inductance is also independent of frequency, and is given by

$$L_e = (\mu/2\pi)\ln(105/30) = 250.6 \text{ nH/m.}$$

The skin depth in copper at 1 MHz is

$$\delta = \frac{1}{\sqrt{\pi \times 10^6 \times 4\pi \times 10^{-7} \times 5.8 \times 10^7}} = 6.608 \times 10^{-5} \text{ m,}$$

In the distributed resistance formula, we have uncanceled dimensions, so we must be sure to convert them to S.I. by multiplying inches by 0.0254 m/in. Also, one must *suspiciously* note that dimensions were given as *diameters*, so a "2" must be added or divided out somewhere. Thus,

$$R_L = \frac{1}{2\pi \times 6.608 \times 10^{-5} \times 5.8 \times 10^7 \times 0.0254} \left(\frac{2}{0.030} + \frac{2}{0.105} \right) = 0.140 \text{ ohms/m}.$$

Now, it is elementary to compute internal inductance by dividing this result by radian frequency: $L_i = 0.14/(2\pi \times 10^6) = 22.2$ nH/m. As this is about 8.9% of the external inductance, the nature of the problem we are solving may govern whether we neglect it or not.

For a dielectric dissipation factor of 2×10^{-4}, the conductivity term becomes

$$G_L = 2\pi \times 10^6 \times 1.004 \times 10^{-10} \times 2 \times 10^{-4} = 1.26 \times 10^{-7} \text{ Siemens/m}$$

EXERCISES 1.2

1. Repeat the calculations in the example above at frequencies of 100 MHz and 10GHz.

ANSWERS Capacitance and external inductance are unchanged.

At 100 MHz, $R_L = 1.4$ ohms/m, internal inductance becomes 2.2 nH/m, so being less than 1% of the external inductance, it is easy to neglect, G_L becomes 1.26×10^{-5}.

At 10 GHz, $R_L = 14.0$ ohms/m, $L_i = 0.22$ nH/m, and $G_L = 1.26 \times 10^{-3}$ S/m.

2. Repeat the calculations of the example above and of Exercise 1 for RG 6U cable, which your friendly Radio Shack store is apt to sell you for connecting TVs or FM to an antenna or to the connector your savvy landlord provided in the living room of your apartment. The main differences between it and the cable in Exercise 1, which is used in a lot of electronics work, are dimensions and dielectrics. RG 6U has 2a = 0.040 inches and 2b = 0.190 inches. The dielectric of RG 6U is foamed polyethylene, used because it reduces attenuation of the cable to practical nonsignificance; its dielectric constant may be used as 1.49, and the author recommends the reader not worry about its dissipation factor nor attempt computation of G_L.

ANSWERS $C_L = 53.2$ pF/m, $L_e = 311.6$ nH/m, $R_L = 0.0989$ ohms/m, and $L_i = 15.75$ nH/m at 1 MHz. $R_L = 0.989$ ohms/m and $L_i = 1.575$ nH/m at 100 MHz. $R_L = 9.89$ ohms/m and $L_i = 0.16$ nH/m at 10 GHz.

3. The beleaguered reader may wonder if there is a yet a practical application of these calculations. The answer is, "yes, partly"; we can see what could be a very bad idea. Suppose we had a video camera with an output impedance of 10 kΩ feeding a VCR with input impedance of 10 kΩ through 5 meters of RG 58U cable. Since video is what we call a "baseband signal," meaning its low frequency is down near zero, the equivalent circuit for the system simply has the two 10 kΩ in parallel with the cable capacitance, which will total around 500 pF. The reader should recognize a low-pass filter, with half-power bandwidth given by B = 1/2πRC. Find B.

ANSWER 636 kHz. Putting the video through such a system may well strip off all the color information, which goes up about to 4 MHz.

4. *The consequences of Exercise 3 illustrate the reason that emitter follower output is such a good idea for devices like the video camera. Suppose the only change you make is to get the camera output impedance down to 100 Ω, which ought not to be difficult at all. Find the new value of low-pass cutoff frequency B.*

ANSWER 31.9 MHz, which ought to be high enough to pass video undiminished.

1.3 Equivalent Circuit for Coaxial Line

In the last section, we found an expression for the capacitance for each meter of coaxial line. The more line there is, the more capacitance; since capacitors add in parallel, all of the capacitors try to be in parallel. Since conductance was calculated directly from capacitance, it also is a parallel element. Inductances add when they are put in series, as do resistors; hence, inductance and resistance are series elements of the equivalent circuit. The result is that the equivalent circuit for a meter or really infinitesimal length of transmission line is as in Figure 1.4.

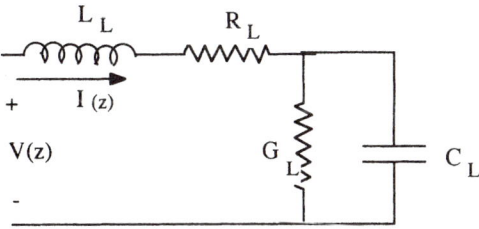

FIGURE 1.4
Equivalent circuit for an incremental length of transmission line.

The first step in analysis of this equivalent circuit is to write equations showing how voltage and current vary along the line. We will make the convenient assumption that we will always be expecting to excite the line sinusoidally, i.e., the form of time expression for voltage would be $v = V(z)e^{j\omega t}$, where we expect the voltage to vary also as a function of z measured along the line, and where our objective is to find out what this variation is. We can expect that the voltage would decrease as we go along the line, due to the voltage drop in L_L and R_L, so we can write

$$-\frac{dV}{dz} = R_L i + L_L \frac{\partial i}{\partial t} = R_L i + j\omega L_L i.$$

We can expect a similar drop in current as we go along the line, due to current being shunted off through C_L and G_L. Hence, a second equation is

$$-\frac{\partial i}{\partial z} = G_L v + C_L \frac{\partial i}{\partial t} = G_L v + j\omega C_L v .$$

Heaviside named these the "telegrapher's equations," not really recognizing that the man with a fast wrist, sitting in a railroad station wearing a green eyeshade, didn't have Heaviside's math skills, but we'll say having a name is better than not having one. For the moment, we will just take a second z-derivative of the first equation and just substitute the second equation for $\frac{di}{dz}$. We get

$$-\frac{d^2v}{dz^2} = \left(R_L + j\omega L_L\right)\frac{di}{dz} = -\left(R_L + j\omega L_L\right)\left(G_L + j\omega C_L\right)v .$$

If we take the second derivative to the other side, we can say we have a second-order differential equation with zero first derivative, so it is of the form

$$\frac{d^2v}{dz^2} - \gamma^2 v = 0 .$$

The solutions to this equation are of the form $Ae^{-\gamma z} + Be^{\gamma z}$, where the quantity

$$\gamma = \sqrt{\left(R_L + j\omega L_L\right)\left(G_L + j\omega C_L\right)} .$$

Here are two complex numbers multiplied, and the square root is taken. If the reader thinks he/she can avoid another complex number, the author also has some swamp land for sale. Typically, we define the parts $\gamma = \alpha + j\beta$. Since the real part α leads to a real exponential in v, we call α the "attenuation constant." In a similar definition, β is called the "phase shift constant." If we so choose, we could also define a "phase velocity" U for waves at this frequency as $U = \omega/\beta$; we would find that at frequencies where internal inductance can be neglected, waves travel at the velocity of light traveling in the dielectric used.

Example 1.3.1

Let us figure attenuation constant and phase shift constant in RG 58U cable at 81 MHz.

SOLUTION Using the numbers from Example 1.2, we first add the internal inductance to the external value and insert numbers into the γ expression. We have

$$\gamma = \sqrt{\left(0.140 + j2\pi \times 10^6 \times 272.6 \times 10^{-9}\right)\left(1.26 \times 10^{-7} + j2\pi \times 10^6 \times 100.4 \times 10^{-12}\right)} .$$

We show a few intermediate results in this calculation to set to rest a few possible impressions of magic upon the reader:

$$\gamma - \sqrt{(0.140 + j1\,713)(1.26 \times 10^{-7} + 6.308 \times 10^{-4})}$$

$$= \sqrt{1.719 \angle 85.33° \times 6.308 \angle 89.99°}$$

$$= \sqrt{1.084 \times 10^{-3} \angle 175.32°} = 3.29 \times 10^{-2} \angle 87.66° - 0.001345 + j0.03289.$$

The units here are nepers/meter for α and radians/meter for β. The consequences are that a wave must travel 743 meters before its amplitude drops to
$1/e$ times the value at the start.
   Phase velocity here is $2\pi \times 10^6 / 0.03289 = 1.91 \times 10^8$ m/s.
   We see that the internal inductance has slowed the wave velocity about
4.5% below the velocity of plane waves in the dielectric,

$$U = \frac{3 \times 10^8}{\sqrt{\varepsilon_r}} = \frac{3 \times 10^8}{\sqrt{2.26}} = 1.996 \times 10^8 \text{ m/s}.$$

EXERCISES 1.3
*1. For RG 58U, find the attenuation constant and phase velocity at 100 MHz and
10 GHz.*
ANSWERS   At 100 MHz, $\alpha = 0.01426$ nepers/m; $U = 1.985 \times 10^8$ m/s.
   At 10 GHz, $\alpha = 0.172$ nepers/m; $U = 1.993 \times 10^8$ m/s.
*2. For RG 6U cable, find the attenuation and phase velocity at 1 MHz, 100 MHz, and
10 GHz.*
ANSWERS   1 MHz: $\alpha = 6.30 \times 10^{-4}$ nepers/m, $U = 2.395 \times 10^8$ m/s
   100 MHz $\alpha = 6.44 \times 10^{-3}$ nepers/m, $U = 2.450 \times 10^8$ m/s
   1 GHz: $\alpha = 6.46 \times 10^{-2}$ nepers/m, $U = 2.455 \times 10^8$ m/s)

1.4   Characteristic Impedance on Coaxial Lines

The ratio between voltage and current for the waves on a transmission line is
defined as the "characteristic impedance." First, we might say how we recognize the individual waves from their mathematical representation. We have
said that the time variation is given by the imaginary exponential $e^{j\omega t}$. We got
two possible solutions to our wave equation, if we write both the real and
imaginary parts of the wave expressions and multiply them by the time variation, the two waves on the line are written mathematically as

$$v = Ae^{j\omega t}e^{-\alpha z}e^{-j\beta z} + Be^{j\omega t}e^{\alpha z}e^{j\beta z}.$$

Now, we can identify which way each wave is traveling by combining the imaginary exponentials. The imaginary exponents with the coefficient "A" are $\omega t - \beta z$, which give the phase variation of that wave. If we set the exponent to some constant, it defines a point of constant phase for the wave. Suppose we pick the constant equal to 0; we would say that at time $t = 0$, that point on the wave is located at $z = 0$. As time goes on, t becomes increasingly positive; so to follow the phase $= 0$ point on the wave, we must let z increase. Hence, this is a wave traveling in the +z direction. Similar reasoning shows that the wave with coefficient "B" is traveling in the –z-direction.

Now, we can say each wave must satisfy the telegrapher's equations. Let us take the positive-going voltage wave and insert it in the first telegrapher's equation. We have

$$-\frac{\partial v}{\partial z} = -e^{j\omega t}A(-\alpha - j\beta)e^{-\alpha z}e^{-j\beta z} = \gamma v_{+wave} = (R_L + j\omega L_L)i_{+wave}.$$

Defining characteristic impedance Z_0 as the ratio of voltage to current for the +wave gives us

$$Z_0 = \frac{v_{+wave}}{i_{+wave}} = \frac{R_L + j\omega L_L}{\gamma} = \frac{R_L + j\omega L_L}{\sqrt{(R_L + j\omega L_L)(G_L + j\omega C_L)}} = \sqrt{\frac{R_L + j\omega L_L}{G_L + j\omega C_L}}.$$

Thus, in general, characteristic impedance is a function of frequency. However, at high frequencies, we can say that $\omega L_L \gg R_L$ and for good dielectrics, always $\omega C_L \gg G_L$. Hence, we may write

$$Z_0 \approx \sqrt{\frac{j\omega L_L}{j\omega C_L}} \approx \sqrt{\frac{L_e}{C_L}}.$$

Indeed, it is a normal approximation that the final expression is stated as the characteristic impedance of a cable.

For the wave going in the –z-direction,

the ratio of v^{-wave} to i^{-wave} is $-Z_0$.

The significance of the negative sign lies in the fact that we define positive current as flowing into the positive voltage conductor; the negative characteristic impedance implies only that that current flows *out* of the positive terminal.

Example 1.4.1

Find the exact characteristic impedance of RG 58U cable at 1 MHz and compare it to the usual approximation.

SOLUTION $\quad Z_0 = \sqrt{\dfrac{0.140 + j2\pi \times 10^6 \times 272.8 \times 10^{-9}}{1.26 \times 10^{-7} + j2\pi \times 10^6 \times 100.4 \times 10^{-12}}}$

$$= \sqrt{1.720\angle 85.33° \; / \; 0.0006308\angle 89.99°} = 52.2\angle - 2.33°$$

This is already very close to the approximate value

$$Z_0 = \sqrt{\frac{250.6 \times 10^{-9}}{100.4 \times 10^{-12}}} = 49.96 \, .$$

In fact, the microwave engineer will invariably refer to RG 58U as an example of "50 ohm cable."

EXERCISES 1.4

1. *Find the exact value of the characteristic impedance of RG 58U cable at frequencies of 100 MHz and 10 GHz.*

ANSWERS $\quad 50.18\angle-0.25°, \; 49.96\angle-0.020$

2. *Find the characteristic impedance of RG 6U cable at frequencies of 1 MHz, 100 MHz, and 10 GHz.*

ANSWERS $\quad 78.49\angle-1.38°, \; 76.72\angle-0.14°, \; 76.56\angle-0.014°.$

   This cable thus has a nominal characteristic impedance of 75 ohms, which is a good match for the folded monopole antenna, or to the 300 ohms of the folded dipole antenna when one also uses the impedance-matching transformer called a balun.

1.5 Terminal Conditions — Reflection Coefficient

On sinusoidally excited lines, the most interesting phenomena occur at or near the load end. Consequently, it is convenient to define waves in terms of their values at the load terminals:

$$Ae^{-\gamma L} \, \Delta V^+ \qquad \text{and} \qquad Be^{\gamma L} \, \Delta V^-,$$

where we may say V is the value at the load of the voltage wave traveling *toward* the load and V⁻ is the value at the load of the wave traveling *away from*

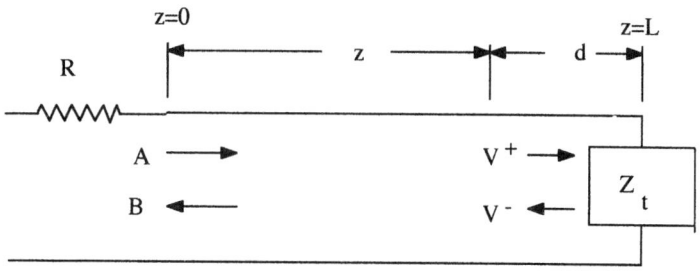

FIGURE 1.5
Directions of voltage waves on transmission line of length L.

the load. Figure 1.5 illustrates where the variables are valid. Similarly, the values of current waves at the load are I^+ and I^-. But since voltage and current in each wave are related through characteristic impedance, we may write

$$I^+ = \frac{V^+}{Z_o} \qquad \text{and} \qquad I^- = -\frac{V^-}{Z_o}.$$

Now the load has a voltage-current relationship given by

$$Z_t = \frac{V_t}{I_t} = \frac{V^+ + V^-}{I^+ + I^-} = \frac{V^+ + V^-}{\dfrac{V^+}{Z_o} - \dfrac{V^-}{Z_o}}.$$

If we now define a reflection coefficient as $\Gamma_T = \dfrac{V^-}{V^+}$ we can obtain its relation to load impedance. We have

$$Z_T = \frac{V^+ + \Gamma_T V^+}{\dfrac{V^+}{Z_o} - \dfrac{\Gamma_T V^+}{Z_o}} = Z_o \frac{1 + \Gamma_T}{1 - \Gamma_T},$$

$$Z_T - \Gamma_T Z_T = Z_o + \Gamma_T Z_o; \; Z_T - Z_o = \Gamma_T (Z_T + Z_o),$$

$$\Gamma_T = \frac{Z_T - Z_o}{Z_T + Z_o} = \frac{\dfrac{Z_T}{Z_o} - 1}{\dfrac{Z_T}{Z_o} + 1}.$$

When we divide an impedance by Z_o, we call the result "normalized impedance." Normalized impedance is often denoted by a lower-case letter; thus,

$$\mathcal{I}_t - \mathcal{I}_r / \mathcal{I}_0 .$$

This will be found to be an especially convenient concept later when we work with the Smith chart.

EXERCISE 1.5

1. Find Γ_T, for $\dfrac{Z_T}{Z_0} = 2, \dfrac{1}{2}, \infty, 0, j, -j$.

ANSWERS $\quad \dfrac{1}{3}, -\dfrac{1}{3}, 1, -1, 1\angle90°, 1\angle-90°$

1.5.1 Generalized Reflection Coefficient

At this point we will define a generalized reflection coefficient, which is the ratio of the two voltage waves at any point on the line. It is particularly useful to handle loss on a transmission line by computing its effect on this generalized coefficient. It is also especially convenient in that modern vector network analyzers commonly read out this variable. Let us rewrite the voltage on the line in terms of the values of the waves at the load end. Solving for A and B, we would get

$$A = V^+ e^{\gamma L} \quad \text{and} \quad B = V^- e^{-\gamma L},$$

$$\text{so } V = V^+ e^{\gamma L} e^{-\gamma z} + V^- e^{-\gamma L} e^{\gamma z} = V^+ e^{\gamma(L-z)} + V^- e^{-\gamma(L-z)}.$$

The quantity $L - z$ measures how far a given point on the transmission line is from the load terminals, and is a useful quantity with which to work in many sinusoidal steady-state problems. We will call it "d" and define the generalized reflection coefficient

$$\Gamma(d) = \frac{V^- e^{-\gamma d}}{V^+ e^{\gamma d}} = \Gamma_T e^{-2\gamma d} = \Gamma_T e^{-2\alpha d} e^{-j2\beta d} .$$

Since Γ_T and γ are both in general complex, we can see that the effect of moving away from the load is to retard the phase and shrink the amplitude of $\Gamma(d)$. The latter effect can be quite salutary in producing a better impedance match at the generator when the load impedance alone would produce a severe mismatch.

Once the generalized reflection coefficient has been defined, it is a big help for future calculations to express the general solutions for voltage and current on the line in terms of $\Gamma(d)$. Factoring $V^+ e^{\gamma d}$ out of the voltage expression, we get

$$V(d) = V^+e^{\gamma d}\left[1 + \frac{V^-e^{-\gamma d}}{V^+e^{\gamma d}}\right] = V^+e^{\gamma d}\left(1 + \Gamma_T e^{-2\gamma d}\right) = V^+e^{\gamma d}\left(1 + \Gamma(d)\right).$$

A similar manipulation can be performed upon the general solution for current on the line, obtaining

$$I(d) = \frac{V^+e^{\gamma d}}{Z_o}\left(1 - \Gamma(d)\right).$$

These expressions for voltage and current are especially useful if one wishes to know what the input impedance of a line would be if it were cut at some distance "d" from the load. One simply divides the voltage expression by the current expression, obtaining

$$Z(d) = \frac{V(d)}{I(d)} = \frac{V^+e^{\gamma d}\left(1 + \Gamma(d)\right)}{\dfrac{V^+e^{\gamma d}}{Z_o}\left(1 - \Gamma(d)\right)} = Z_o\,\frac{1 + \Gamma(d)}{1 - \Gamma(d)}.$$

Example 1.5.1

Given $\Gamma_T = 0.3\angle 45°$, what will be the reflection coefficient at the input end of a transmission line 0.3 λ long with a one-way attenuation of 2 dB? If the characteristic impedance of the line is 50 ohms, what termination impedance caused the reflection coefficient and what is the input impedance of the line?

SOLUTION We need to evaluate $\Gamma_T e^{-2\alpha L}e^{-j2\beta L}$. Since $e^{-\alpha L}$ tells how much a voltage wave is attenuated in traveling one way on the line, we must say

$$-2\text{ dB} = 20\log_{10}e^{-\alpha L} = 10\log_{10}e^{-2\alpha L}, \text{ and}$$

$$e^{-2\alpha L} = 10^{-0.2} = 0.631.$$

A more convenient way of expressing β is $\dfrac{2\pi}{\lambda}$. Then,

$$2\beta L = 2\left(\frac{2\pi}{\lambda}\right)(0.3\lambda) = 1.2\pi = 216°$$

$$\Gamma(L = 0.3\lambda) = 0.3 \times 0.631\angle 45° - 216° = 0.189\angle -189°.$$

The expression for Z(d) may be evaluated for any value of "d," including zero. Hence,

$$Z_T = Z_n \frac{1+\Gamma_T}{1-\Gamma_T} = 50 \frac{1+0.3\angle45°}{1-0.3\angle15°} = 50 \frac{1+0.2121+j0.2121}{1-0.2121-j0.2121}$$

$$= 50 \frac{1.2306\angle9.93°}{0.0130\angle-13.00°} = 75.41\angle25.00° = 68.35 + j31.86 \ \Omega.$$

Similarly,

$$Z(d) = Z_o \frac{1+\Gamma(d)}{1-\Gamma(d)} = 50 \frac{1+0.189\angle-171°}{1-0.189\angle-171°}$$

$$= 50 \frac{1-0.1870-j0.0296}{1.1870+j0.0296} = 34.26\angle-3.51° = 34.20 - j2.10 \ \Omega.$$

EXERCISES 1.5

1. Given $\Gamma_T = 0.8\angle30°$, what is $\Gamma(L)$ $4\frac{1}{4}\lambda$ away from the load on a line on which the one-way attenuation is 3 dB? If the characteristic impedance of the line is 75 ohms, what termination impedance caused the reflection coefficient and what is the input impedance of the line?

ANSWERS $0.4\angle210°$, $Z_T = 106.15 + j235.9 \ \Omega$, $Z(d) = 34.00 - j\,16.19 \ \Omega$

2. The effect of attenuation on the generalized reflection coefficient suggests an effective method for determining the attenuation on a line. In principle, one connects a known reflection coefficient to the load terminals and measures the generalized reflection coefficient at the other end. In practice, the most convenient reflection coefficient to provide is −1, corresponding to a short circuit. Suppose a 50 ohm line is shorted and the input impedance is found to be 350 ohms. Find the one-way attenuation of the line as a decimal fraction and in deciBels.

ANSWER $e^{-\alpha L} = 0.866 \rightarrow -1.25$ dB

3. Repeat 2 for a shorted 50 ohm line with an input impedance of 3 ohms.

ANSWER $e^{-2\alpha L} = 0.887 \rightarrow -0.52$ dB

4. Suppose the input impedance is 5 ohms when a certain 75 ohm line is shorted. What is its attenuation factor?

ANSWER $e^{-\alpha L} = 0.935 \rightarrow -0.58$ dB

1.6 Mr. Smith's Invaluable Chart

In many problems as in Exercise 4 above, we see that neglecting attenuation totally would not greatly affect the calculated results. Even before World War II, a Bell Labs scientist named P. H. Smith derived a graphical aid to calculation which greatly simplified the calculation of input impedance, especially for lossless lines. On lossless lines, the generalized reflection coefficient keeps a constant amplitude, varying only in phase as one travels along the line; hence, the locus of generalized reflection coefficient is a circle of radius determined by the amplitude of the coefficient. Smith might have said, "Consider generalized reflection coefficient to be a complex variable with magnitude less than or equal to unity. We will write the variable as u + jv." Also, let us divide the input impedance by characteristic impedance Z_0. The result is called "normalized impedance;" from here forward, we will use lower-case letters for normalized impedance or admittance. Hence, z = r + jx. Thus, we write normalized impedance as

$$Z(d)/Z_0 = r + jx = \frac{1+u+jv}{1-(u+jv)}.$$

We need to separate the right-hand side into real and imaginary parts, which we will equate to r and x. This is done by multiplying by the complex conjugate of the denominator. Thus,

$$r + jx = \frac{1+u+jv}{1-u-jv} \times \frac{1-u+jv}{1-u+jv} = \frac{1-u^2-v^2+j2v}{(1-u)^2+v^2} ; \text{ hence}$$

$$r = \frac{1-u^2-v^2}{u^2-2u+1+v^2} ; \ u^2 r - 2ur + r + v^2 r = 1 - u^2 - v^2.$$

Now, the operation we need to do is called "completing the square" in u and v. First we move several terms, getting

$$u^2(r+1) - 2ur + v^2(r+1) = 1 - r,$$

$$u^2 - \frac{2ur}{r+1} + v^2 = \frac{1-r}{1+r}.$$

Here the square is already completed in v, but the second term has u times "something," and we must simply add the square of one half the "something," which is r/(r+1), to both sides of the equation. So,

$$u^2 - \frac{2ur}{r+1} + \left(\frac{r}{r+1}\right)^2 + v^2 = \left(u - \frac{r}{r+1}\right)^2 + v^2 = \frac{1-r}{1+r} + \left(\frac{r}{r+1}\right)^2$$

$$= \frac{1-r^2+r^2}{(1+r)^2} = \left(\frac{1}{r+1}\right)^2.$$

Now, our long-forgotten knowledge of analytic geometry will tell us that we have a family of circles, with centers at $u = r/(r+1)$ and $v = 0$, of radius $1/(r+1)$. Let us just do several of these:

a. For $r = 0$, the circle is centered at $u = 0$, $v = 0$, i.e., the origin, of radius 1, that is a circle of unit radius centered at the origin.

b. For $r = 1$, the center is at $1/2$, 0, and of radius $1/2$. For $r = 3$, the center is at $3/4$, 0, and of radius $1/4$.

It can be seen that all r-circles go through 1, 0.

The equation for x almost has its square complete as we start

$$x = \frac{2v}{(1-u)^2 + v^2}; \text{ hence, } (u-1)^2 + v^2 = \frac{2v}{x}.$$

The square is completed in u, and we complete it in v by bringing $2v/x$ onto the left-hand side and adding $(1/x)^2$ to both sides, obtaining

$$(u-1)^2 + \left(v - \frac{1}{x}\right)^2 = \left(\frac{1}{x}\right)^2.$$

Thus, circles of constant x are centered at $u = 1$, $v = 1/x$, and have radius $1/x$. On our sketch, which we call Figure 1.6, we have shown circles for $x = 0$, $x = \pm 1/2$ and ± 1. Copies of the chart with many circles of both r and x are available at a nominal price in the bookstores of most colleges offering engineering or electronics technology; they may also, at this writing, be copied off the Internet.

Example 1.6

Once again, if Candide were running the world, Smith charts would be perfectly round and precise in all ways. The author's experience is that some unnamed villains are subjecting good, accurate Smith charts to photocopy machines, which have a slight tendency to stretch the image a bit vertically, so that what was once perfectly round is now slightly elliptical. We will act like Candide, and assume that the $|\Gamma| = 1$ ($r = 0$) circle on the Smith chart is precisely 91.2 mm in radius, but will give answers not from graphical calculations but from numerical ones, as demonstrated in Appendix A.

Suppose we wish to know what normalized impedance corresponds to a reflection coefficient of $1/3$ at $0°$.

SOLUTION The magnitude of this reflection coefficient will be given by one third of 91.2 mm, or 30.4 mm. From the center of the chart, we measure 30.4 mm horizontally right, and if we have a perfectly round chart, we should find $z = 2.0 + j0$.

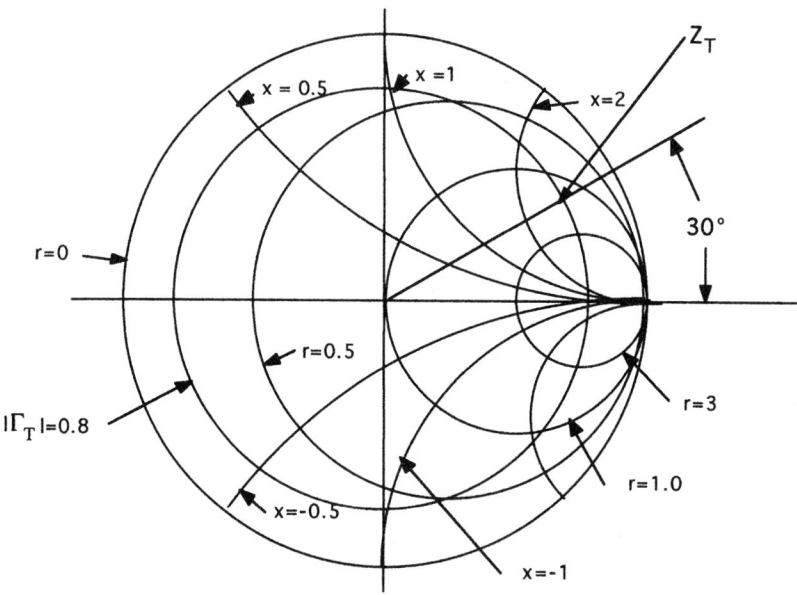

FIGURE 1.6
Smith chart constructed from discussion in text.

EXERCISES 1.6

1. *Using the Smith chart, find the normalized impedance corresponding to 0.333 at 90°, 180°, and 270°.*

ANSWERS $0.8 + j0.6$, 0.5, $0.8 - j0.6$.

2. *Find the normalized impedance corresponding to reflection coefficient magnitude of 0.5 at angles of 0°, 90°, 180°, and 270°.*

ANSWERS 3.0, $0.6 + j0.8$, 0.33, $0.6 - j0.8$

3. *Find the normalized impedance corresponding to reflection coefficients of $0.2\angle 30°$, $0.3\angle{-}40°$, $0.4\angle{-}110°$, $0.5\angle{-}270°$.*

ANSWERS $1.38 + j0.29$, $1.44 - j0.61$, $0.59 - j0.52$, $0.6 + j0.8$.

4. *Find the reflection coefficients if the following impedances are connected to a 50 ohm line: $10 + j20$, $75 - j25$, $100 - j200$, $- j50$, $j100$.*

ANSWERS $0.71\angle 135°$, $0.28\angle{-}33.7°$, $0.82\angle{-}22.8°$, $1\angle{-}90°$, $1\angle 53.1°$.

1.7   Characteristic Impedance of Microstrip Line

We did a distributed circuit analysis of coaxial line because (1) historically, it has been very important to RF engineers, and (2) its simple geometry leads to fairly simple mathematics. Another kind of line has become fairly common

Because it connects rather conveniently to integrated circuit board electronics. It is called "microstrip" and its geometry is as shown in Figure 1.7.

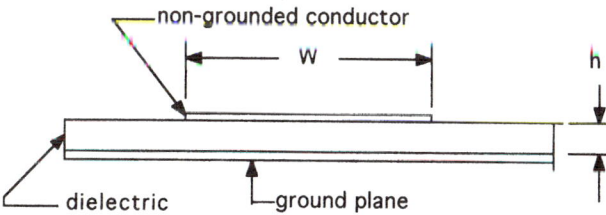

FIGURE 1.7
Geometry of microstrip circuit.

The analysis of microstrip can be simple, but only when its dimensions permit neglecting fringing of electric fields outside the region between the plates, that is, when spacing "h" is small compared to width "W" of the signal conductor. Unfortunately, such conditions would limit one to rather low characteristic impedances, and today's microwave engineer requires the flexibility of a variety of characteristic impedances. Our procedure will be to perform the approximate analysis which is accurate when width is large compared to spacing, and then for narrow conductor width, will simply use the fairly messy formula which a more precise analysis yields.

When the approximation $W/h \gg 1$ is valid and fringing of electric fields can be neglected, we can recall elementary physics and simply write the capacitance per meter of distance into or out of the page as

$$C_L = \frac{\varepsilon_r \varepsilon_0 W}{h} \text{ Farads/meter.}$$

Next, we recall the results of coaxial cable analysis, in which it was found that at high frequencies, inductance is mainly the *external* inductance, which is very reminiscent of the capacitance formula, with ε replaced by μ, and reciprocals of all dimensions, thus

$$L_L \approx L_e = \frac{\mu_r \mu_0 h}{W} \text{ Henries/meter.}$$

μ_r is the relative permeability of the dielectric and is close to unity for practical dielectrics. Thus, when $W/h \gg 1$, high frequency characteristic impedance is well approximated as

$$Z_{||} = \sqrt{\frac{L_e}{C_L}} \approx \sqrt{\frac{\mu_0 h}{W} \times \frac{h}{\varepsilon_r \varepsilon_0 W}} = \frac{1}{\sqrt{\varepsilon_r}} \sqrt{\frac{\mu_0}{\varepsilon_{||}} \frac{h}{W}}.$$

Suppose we evaluate this impedance for $\varepsilon_r = 2.25$, which is close to the value for one grade of "Duroid," a product of Rogers Corporation, of Chandler, AZ, and for $W = 10h$. We note that the quantity $\sqrt{\dfrac{\mu_0}{\varepsilon_0}}$ is known in wave theory as the "intrinsic impedance of vacuum," and is approximated by 120π ohms. Thus, characteristic impedance is

$$Z_0 \approx \sqrt{\frac{1}{2.25}} \times 120\pi \times \frac{1}{10} = 8\pi = \text{about 25 ohms.}$$

The exact solution for width of the conductor versus spacing for a given design value of characteristic impedance is complicated by being a "two-dielectric problem." The definitive word comes from Harold Wheeler in an IEEE article published in 1977, the year in which he turned 74 years of age. This writer judged the more useful result to be W/h for a specified characteristic impedance. If it is assumed that the thickness of the conductor is also small compared with the spacing, the formula becomes[*]

$$\frac{W}{h} = 8 \frac{\sqrt{\left[\exp \dfrac{Z_0}{42.4}\sqrt{\varepsilon_r + 1} - 1\right]\dfrac{7 + 4\,\varepsilon_r}{11} + \dfrac{1 + 1/\varepsilon_r}{0.81}}}{\left[\exp \dfrac{Z_0}{42.4}\sqrt{\varepsilon_r + 1} - 1\right]}.$$

Another consequence of having two different dielectrics is that waves travel on the line at a speed which is a complicated function of the dielectric constant and of the ratio W/h. The speed of waves is given by $\dfrac{c}{\sqrt{\varepsilon_{eff}}}$, where c is the speed of plane waves in vacuum and ε_{eff}, the effective value of the dielectric constant, is given by[**]

$$\varepsilon_{eff} = \frac{\varepsilon_r + 1}{2} + \frac{\varepsilon_r - 1}{2}\left(1 + \frac{12h}{W}\right)^{-1/2}.$$

Example 1.7
Let us suppose that we have microstrip material in which the actual dielectric constant is $\varepsilon_r = 2.25$, the thickness h of the dielectric is 1.0 mm. For a characteristic impedance of 50 ohms, we wish to calculate the width W of the non-grounded conductor and the wavelength at frequency of 1000 MHz (1.0 GHz).

[*] Wheeler, H.A., "Transmission-Line of a Strip on a Dielectric Sheet on a Plane," *IEEE Transactions of Microwave Theory and Techniques*, Vol. 25, pp. 631–647, 1977.
[**] Posar, David M., *Microwave Engineering, 2nd Edition*, John Wiley & Sons, New York, 1998, p. 102.

(The impedance matching techniques we will study later involve using lengths of transmission lines which must be expressed in terms of wavelength.)

SOLUTION $\quad W/h = 8 \dfrac{\sqrt{\left[\exp\dfrac{50}{42.4}\sqrt{2.25+1}-1\right]\dfrac{7+4\,2.25}{11}+\dfrac{1+1/2.25}{.81}}}{\left[\exp\dfrac{50}{42.4}\sqrt{2.25+1}-1\right]}$

If the reader wishes to check intermediate results, the author's result for the denominator (which also appears as a multiplier in the numerator) is 7.380, and the overall result is 3.00, or W = 3 mm.
Then,

$$\varepsilon_{eff} = \frac{2.25+1}{2} + \frac{2.25-1}{2}\left(1+12/3.00\right)^{-1/2} = 1.905.$$

Thus, the wavelength, which would be 30 cm in vacuum and 20 cm in a one-dielectric system completely insulated by this dielectric, would here be

$$\lambda = \frac{30}{\sqrt{1.905}} \text{ cm} = 21.74 \text{ cm}.$$

EXERCISES 1.7
1. For dielectrics of dielectric constant 3.27, 4.5, 6.15, and 9.8, find the value of W/h for a 50 ohm characteristic impedance, the effective dielectric constant, and the wavelength at 1 GHz.

ANSWERS   They are given in the table below, as a lesson to computer-phobics like the author. Even for us, such elementary ideas as spreadsheets can give superior results, in far less time, than the most powerful calculators, unless perhaps the latter are *also* programmed.

ε_r	[ ] in formula	W/h	ε_{eff}	λ
3.27	10.436	2.352	2.594	18.625
4.5	14.888	1.876	3.393	16.286
6.15	22.411	1.473	4.426	14.259
9.8	47.202	0.976	6.607	11.671

1.8   Two-Port High Frequency Models for Bipolar Transistors

Normally, a student's introduction to high frequency behavior may begin with the "hybrid-pi high frequency equivalent circuit" for the bipolar transistor, as shown in Figure 1.8.

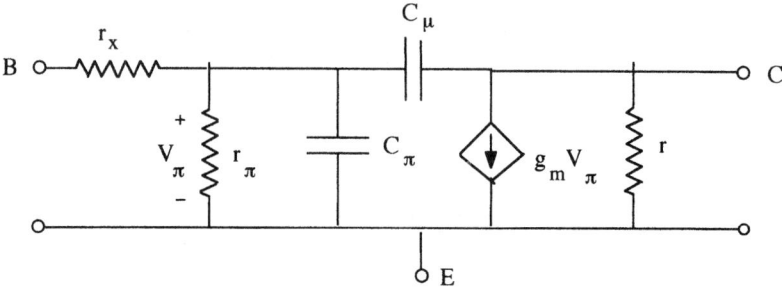

FIGURE 1.8
Hybrid-pi high frequency model for bipolar transistor.

The capacitors shown cause the gain of the transistor to drop off at high frequencies, for at least two different reasons. First, both capacitors lower the input impedance as the frequency increases; hence, more of the input voltage is lost across the "base spreading resistor" r_x. Second, the capacitor connected across the top, designated C_μ, effectively sends output voltage back to the input; but since there is a phase inversion from input to output, we have provided *negative* feedback, which also decreases gain.

The usual two-port models used in junior level electronics books are the so-called "hybrid parameters;" their hybridness has to do with the fact that no two of them have the same dimensions. The defining equations are

$$V_1 = h_{11}I_1 + h_{12}V_2,$$

$$I_2 = h_{21}I_1 + h_{22}V_2.$$

Some of the fundamental definitions designate that V_1 appears between the left-hand terminals of the device, with positive terminal on top, and I_1 is positive when flowing into the top left terminal; V_2 appears between the right-hand terminals, with again the top one being the positive one, and with I_2 being positive flowing into the top right terminal. The equations themselves give us clues as to how to determine the various hybrid parameters. If we make $V_2 = 0$, then h_{11} will be the input impedance of the transistor, and h_{21} will be a current gain from input to output. We of course make $V_2 = 0$ by short-circuiting terminals 2. Continuing the procedure, we would want to make $I_1 = 0$, which we would do by connecting *nothing* to the input terminals. Then h_{22} is an output *admittance* and h_{12} is a voltage gain from output to input.

While the determination of hybrid parameters seems straightforward enough, a very significant problem arose when it was attempted with the early high-frequency bipolar transistors. They tended to become unstable, which is to say, they gave an output with no input, at certain frequencies, when they were either shorted or open-circuited. So, some unnamed genius among microwave engineers said, "Well, if both high and low impedances are bad, we had better try some moderate values, and because there is so much 50 ohm coaxial cable in the world, let's measure characteristics with 50 ohm generator and output loads, and transmission lines connected to the transistor having characteristic impedance of 50 ohms." There have been various stages in the definition of these parameters, but for our use, it is sufficient to let the letter "a" represent the voltage phasor amplitude of a wave headed *towards* a set of transistor terminals and "b" is a phasor value for a wave which is traveling *away* from terminals. Then, the defining equations become

$$b_1 = S_{11}a_1 + S_{12}a_2,$$

$$b_2 = S_{21}a_1 + S_{22}a_2.$$

Now, we need to think in terms of transmission line theory to understand how we make a_1 or $a_2 = 0$. If we would put a generator in the input circuit and consider the transistor to be the load for that transmission line, then a_2 would represent a wave flowing from the load towards the output terminal. Clearly, we could "kill" a_2 by simply connecting a 50 ohm load to the output terminals. Under those conditions, $S_{11} = b_1/a_1$, and is simply the input reflection coefficient with 50 ohms connected to the output terminals. $S_{21} = b_2/a_1$ is simply a transmission coefficient from input to output, under these conditions. Because a major reason for using a transistor is to obtain some amplification, we will usually hope that $|S_{21}| \gg 1$. Note that we are thoroughly into complex number territory here, so any of these parameters, which incidentally are called "scattering coefficients" (probably because the first big use of microwaves during World War II was for radar, and these waves reminded engineers of the reflections from radar targets) and also will have phase angles, are usually quoted between $+180°$ and $-180°$. The procedure for determining the other two scattering coefficients is similar; this time, we put the generator in the output circuit, whereupon S_{22} is the output reflection coefficient with a 50 ohm load connected at the input, and S_{12} is a transmission coefficient from output back to input.

The HP8753 Network Analyzer will graph each scattering parameter over the range of frequency chosen. If one needs the parameters at certain specific frequencies, one can put the "marker" on each frequency desired and the magnitude and phase of the complex number will be read out. Motorola provides data in both formats; some data copied from *Motorola RF Device Data* are also graphed in the complex plane, as shown in Figure 1.9.

Example 1.8

From the scattering coefficients, find the input and output impedance when the other terminals are terminated in a 50 ohm load, for the MRF571 at 200 MHz.

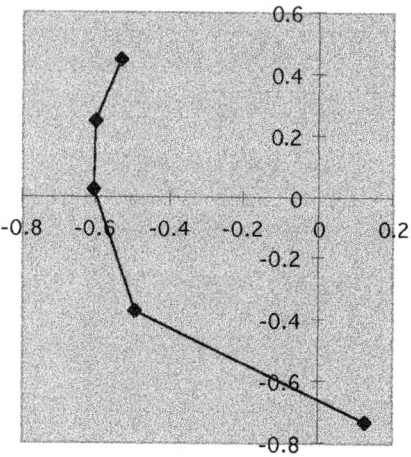

FIGURE 1.9
A plot of parameter S_{11} for several frequencies.

SOLUTION   We may consider $S_{11} = 0.74\angle-86°$ and $S_{22} = 0.69\angle-42°$ to be a load reflection coefficient at one end of the transistor when the other end is loaded by 50 ohms real. Hence,

$$Z_{in} = Z_0 \frac{1+S_{11}}{1-S_{11}} = 50 \frac{1+0.74\angle-86°}{1-0.74\angle-86°} = 15.6 = j51.11 \text{ ohms, and}$$

$$Z_{out} = Z_0 \frac{1+S_{22}}{1-S_{22}} = 50 \frac{1+0.69\angle-42°}{1-0.69\angle-42°} = 50.14 = j102.47 \text{ ohms.}$$

1.9   Input and Output Impedances: General Case

Just above, it was slightly artificial to obtain input and output impedances by terminating the terminals not being fed by the generator with Z_0; as we look at the defining equations for S-parameters, this made $a_2 = 0$, so that the input reflection coefficient became S_{11}. Looking still at the equation for b_1, we see that we would also obtain

$$\Gamma_{in} = S_{11} \text{ if } S_{12} \to 0.$$

This would mean that the transistor was what is called "unilateral," i.e., signal flows through it in only one direction. If one compares the values for the scattering coefficients at various frequencies, it is noticeable that S_{12} increases as frequency increases. This should not be too surprising, as it arises from capacitance, which in the bipolar transistor is C_μ and, of course, capacitive admittance increases with frequency, allowing more and more signal to be fed from the output circuit back to the input. However, one may require an expression for input reflection coefficient which is valid for all conditions. Suppose we relate a_2 and b_2 through a reflection coefficient; to wit,

$$a_2 = \Gamma_L b_2.$$

We substitute thus for b_2 in the second defining equation

$$\frac{a_2}{\Gamma_L} = S_{21}a_1 + S_{22}a_2.$$

This needs to be solved for a_2 and the result substituted in the other defining equation

$$a_2\left(\frac{1}{\Gamma_L} - S_{22}\right) = \left(\frac{1 - S_{22}\Gamma_L}{\Gamma_L}\right)a_2 = S_{21}a_1; \ a_2 = \frac{S_{21}\Gamma_L a_1}{1 - S_{22}\Gamma_L}.$$

Now, substituting in the first defining equation,

$$b_1 = S_{11}a_1 + \frac{S_{12}S_{21}\Gamma_L a_1}{1 - \Gamma_L S_{22}}.$$

Next, defining the input reflection coefficient

$$\Gamma_{in} = \frac{b_1}{a_1} = S_{11} + \frac{S_{21}S_{12}\Gamma_L}{1 - \Gamma_L S_{22}},$$

one can obtain an absolutely analogous expression for Γ_{out}, postulating that the input terminals will be terminated with a reflection coefficient,

$$\Gamma_G = a_1/b_1,$$

obtaining, through similar complex algebra,

$$\Gamma_{out} = S_{22} + \frac{S_{21}S_{12}\Gamma_G}{1 - S_{11}\Gamma_G}.$$

If one further examines the scattering coefficients as a frequency increases, one sees that the amplifying coefficient S_{21} approaches unity or less. Thus, the gain of a stage could be small indeed, if care is not taken to match impedances as much as possible. Indeed, the best one could possibly do would be to have both input and output circuits simultaneously conjugate-matched, which is to say, at both input and output of the transistor, the input impedance is the complex conjugate of the generator impedance, and the output impedance is the complex conjugate of the transistor output impedance. This author has not been able to track down the unrecognized engineer who first slogged through a jungle of complex algebra to derive the conditions for simultaneous conjugate match. If the reader wishes to expose him/herself to this derivation before taking advantage of it, Pozar* gives a fairly thorough modern treatment. It should be noted, at the outset, that it is only possible to have input and output simultaneously conjugate-matched if the transistor is *unconditionally* stable at that frequency. Pozar and others provide us with the necessary conditions for unconditional stability. First, the determinant of the scattering matrix must have a magnitude less than unity. Defining the determinant as Δ, then $\Delta = S_{11}S_{22} - S_{21}S_{12}$. Note that nowhere in this expression have we written a simple magnitude, but the first stability criterion is $|\Delta| < 1$. The other necessary stability criterion is $|K| > 1$, where K is defined as

$$K = \frac{1 - |S_{11}|^2 - |S_{22}|^2 + |\Delta|^2}{2|S_{12}S_{21}|}.$$

If both these conditions are met, one can calculate the value of Γ_L and Γ_G such that there will be simultaneous conjugate matches at both sets of terminals. There are some intermediate results which make the formulas at least look more approachable. They are

$$B_1 = 1 + |S_{11}|^2 - |S_{22}|^2 - |\Delta|^2,$$

$$B_2 = 1 - |S_{11}|^2 + |S_{22}|^2 - |\Delta|^2,$$

$$C_1 = S_{11} - \Delta S^*_{22},$$

$$C_2 = S_{22} - \Delta S^*_{11}.$$

Then, the load reflection coefficient for conjugate matching at both sets of terminals is

$$\Gamma_{LM} = \frac{B_2 - \sqrt{B_2^2 - 4|C_2|^2}}{2C_2}.$$

* David M. Pozar, *Microwave Engineering*, Second Edition, John Wiley & Sons, New York, 1998, Chap. 11.

The comparable reflection coefficient at the input end is

$$\Gamma_{OM} = \frac{B_1 - \sqrt{B_1^2 - 4|C_1|^2}}{2C_1}.$$

The transducer gain we may expect if we meet all these conditions is then given by

$$G_{TM} = (|S_{21}|/|S_{12}|) \times (K - \sqrt{K^2 - 1}).$$

Example 1.9

Let us now compute all these numbers for the MRF 571 transistor, at $I_C = 5.0$ mA, $V_{CE} = 6.0$ V, and at a frequency of 1 GHz.

SOLUTION   For the conditions stated, we can read $S_{11} = 0.61\angle178°$, $S_{21} = 3.0\angle78°$, $S_{12} = 0.09\angle37°$, and $S_{22} = 0.28\angle-69°$.
Hence,

$$\Delta = (0.61\angle178°)(0.28\angle-69°) - (3.0\angle79°)(0.09\angle37°) = (0.1017\angle-54.89°).$$

Then,

$$K = \frac{1 - (0.61)^2 - (0.28)^2 + (0.1017)^2}{2(3.0)(0.09)} = 1.0367.$$

Thus, we have established that the transistor *is* unconditionally stable at this frequency, so we may proceed to find the reflection coefficients for simultaneous conjugate match.

$$B_1 = 1 + (0.61)^2 - (0.28)^2 - (0.1017)^2 = 1.2833$$

$$B_2 = 1 + (0.29)^2 - (0.61)^2 - (0.1017)^2 = 0.7017$$

$$C_1 = 0.61\angle178° - (0.1017\angle-54.89°)(0.28\angle+69°) = 0.6374\angle178.71°$$

$$C_2 = 0.28\angle-69° - (0.1017\angle-54.89°)(0.61\angle-178°) = 0.3400\angle-66.10°$$

At the input, the magnitude of the reflection coefficient for simultaneous match is

$$|\Gamma_{MG}| = \frac{1.2783 - \sqrt{1.2783^2 - 4(.6374)^2}}{2 \times 0.6374} = 0.890.$$

Now, because C_1 appears in the denominator as the only complex number in the equation, the phase of Γ_{MG} is the negative of the phase of C_1; therefore,

$$\Gamma_{MG} = 0.890\angle -178.71°,$$

$$|\Gamma_{MG}| = \frac{0.7017 - \sqrt{(0.7017)^2 - 4(0.34)^2}}{2 \times (0.34)} = 0.777;$$

its phase is of course the negative of the phase of C_2, or $66.10°$. The gain which may be expected if these reflection coefficients are used is

$$G_{TM} = \frac{3}{0.09}\left(1.0367 - \sqrt{(1.0367)^2 - 1}\right) = 25.44, \text{ representing 14.06 dB.}$$

EXERCISES 1.9

1. Demonstrate that conjugate match has been obtained in the example above, i.e., let $\Gamma_L = \Gamma_{LM}$ and calculate Γ_{in}, hoping to get the conjugate of Γ_{GM}. Also, let $\Gamma_G = \Gamma_{GM}$ and calculate Γ_{out}, hoping to get the conjugate of Γ_{ML}.

ANSWERS   They come out really close to the correct results. Repeat either or both until you get them right.

2. Perform all the calculations of the example above for the MRF 571 transistor at 1500 MHz.

ANSWERS $\Delta = 0.1296\angle -23.97°$, $K = 1.197$.
   Hence, unconditionally stable at this frequency.
   $\Gamma_{MG} = 0.803\angle -161°,$
   $\Gamma_{ML} = 0.6075\angle 70.45°,$
   $G_{TM} = 9.8 \rightarrow 9.9$ dB.

1.10 Stability Circles for the Potentially Unstable Transistor

The author has forgotten which old Woody Allen movie it was where he and the lady were prowling around a scary environment when Woody says, "This is no time for panic." Five seconds later, they find a scary thing and he says, "Now is the time for panic!" as they begin their flight. Similarly, the microwave engineer might panic prematurely when he/she finds frequencies where a transistor is not unconditionally stable. Fortunately, geometric constructions called "stability circles," when drawn upon a Smith chart, are useful in showing what values of input and/or output reflection coefficients must be avoided to keep the transistor stable. The circle which applies at the transistor input[*] is centered at

$$C_G = \frac{\left(S_{11} - \Delta S_{22}^*\right)^*}{|S_{11}|^2 - |\Delta|^2},$$

and has a radius given by

$$R_G = \left|\frac{S_{12}S_{21}}{|S_{11}|^2 - |\Delta|^2}\right|,$$

where this author writes "G" for the generator end rather than Pozar's "S" for "sending." At the load end, we have

$$C_L = \frac{\left(S_{22} - \Delta S_{11}^*\right)^*}{|S_{22}|^2 - |\Delta|^2} \quad \text{and} \quad R_L = \left|\frac{S_{12}S_{21}}{|S_{22}|^2 - |\Delta|^2}\right|.$$

It turns out that the MRF 571 is potentially unstable at 500 MHz. Let us find the warning phenomena and then find the stability circles to find regions of safety in the following example.

Example 1.10
Compute Δ and K for the MRF 571 transistor at 500 MHz to find the potential instability. Then calculate center and radius of each stability circle and sketch them onto a Smith chart.

SOLUTION At 500 MHz, we have $S_{11} = 0.62\angle-143°$, $S_{21} = 5.5\angle97°$, $S_{12} = 0.08\angle33°$, and $S_{22} = 0.41\angle-59°$.

$$\Delta = (0.62\angle-143°) \times (0.41\angle-59°) - (5.5\angle97°) \times (0.08\angle33°) = 0.2464\angle-78.97°$$

Because $|\Delta| < 1$, one of the necessary conditions for unconditional stability is met.

$$K = \frac{1 + (0.2464)^2 - (0.62)^2 - (0.41)^2}{2 \times 5.5 \times 0.08} = 0.5775.$$

This latter result is not greater than one, so the transistor fails one of the necessary conditions for unconditional stability. However, instead of simply throwing up our hands, we draw stability circles.

---

[*] Pozar, op. cit.

$$C_G = \frac{\left(0.62\angle -143° - (0.2464\angle -78.97°)(0.41\angle 59°)\right)^*}{(0.62)^2 - (0.2464)^2}$$

$$= 2.102\angle 150.15° = -1.82 + j1.05$$

$$R_G = \frac{5.5 \times 0.08}{(.62)^2 - (.2464)^2} = 1.36$$

$$C_L = \frac{\left(0.41\angle -59° - (0.2464\angle -78.97°)(0.62\angle 143°)\right)^*}{(0.41)^2 - (0.2464)^2} = 4.75\angle 73.55° = 1.34 + j4.55$$

$$R_L = \frac{5.5 \times 0.08}{(0.41)^2 - (0.2464)^2} = 4.1$$

Parts of the stability circles are sketched in Figure 1.10, where the origin is the center of the Smith chart.

EXERCISES 1.10
1. *Repeat the exercise above at 200 MHz, where*

$$S_{11} = 0.74\angle 86°,$$

$$S_{21} = 10.5\angle 129°,$$

$$S_{12} = 0.06\angle 46°,$$

$$S_{22} = 0.69\angle -42°.$$

ANSWERS $\Delta = 0.5543\angle -55.89°$, $K = 0.225$ (potentially unstable), $C_G = 1.33 + j2.70$, $R_G = 2.62$, $C_L = 0.94 + j3.96$, $R_L = 3.73$.

2. *The consequence of using a reflection coefficient in the "forbidden" areas defined by the stability circles is that input or output reflection coefficients are greater than one, meaning more energy comes out than goes in. From the example above, check input reflection coefficient for* $\Gamma_L = 0.8\angle 70°$, *and output reflection coefficient for* $\Gamma_G = 0.9\angle 150°$.

ANSWERS $\Gamma_{in} = 1.128\angle -148.46°$, $1.281\angle -67.41°$.

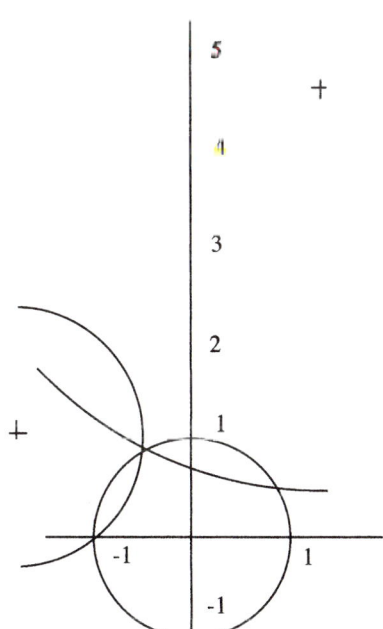

FIGURE 1.10
Smith chart unit circle minus unstable regions.

2

Impedance Matching Techniques

2.1 Introduction

We saw in the previous chapter that optimizing the performance of high frequency transistors may require the connection of rather arbitrary impedances between generator and transistor input, and between the output and the eventual 50 ohm load impedance. There are a number of techniques now available to the high frequency circuit designer, and making a choice may depend upon whether the frequency of operation is VHF, UHF, or microwave, how broad a band of frequencies must be handled by the design, availability of "chip" components of appropriate physical size and circuit value, and other problems. When microwave circuits were just coming into use, from World War II onward, the main method was first to compute the distance from the load at which a purely shunt reactive circuit element would perfectly match the transmission line. Then the reactance was fabricated as a "stub," which was another length of transmission line with a good solid short as its load, or in waveguides, as a kind of window in which a thin slice of metal closed off part of the waveguide. In both of these cases, the reactance shunted the line, that being the only physically feasible connection for coaxial cable and waveguide. More recently, widespread use of microstrip lines, in which the signal-bearing conductor is very accessible, have made it very reasonable to connect a "surface mount" capacitor or inductor in *series* with the line, so more flexibility is available to the circuit designer. We will start with a technique that should be entirely understandable to the electrical engineering student as soon as ac circuit analysis has become a useful part of his/her analysis tools.

2.1.1 Matching Impedances Using Reactive L-Sections

Let us hasten to state that by an "L-section" we mean not only inductances, but that the circuit diagram will have reactive elements drawn at right angles to each other, as in the capital letter "L." Figure 2.1 shows two basic connections for these two "Ls."

In each of the cases diagrammed, we have a resistive load which we wish to modify using the matching section to match to a different real impedance.

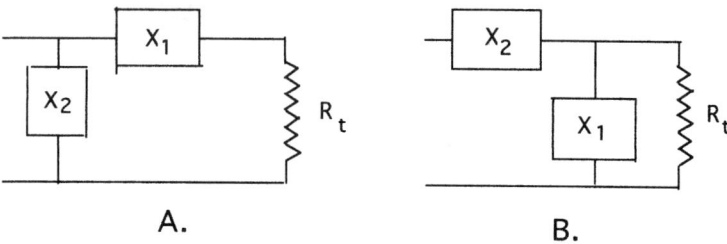

FIGURE 2.1
Two possible orientations of L-sections to achieve impedance match: A. Real load impedance less than resistance to be matched to; B. Real load impedance greater than resistance to be matched to.

We will find that the first diagram has the effect of matching the load to a larger impedance. The analysis and design is straightforward and quick.

2.1.2   Matching Lower Impedance to Higher Impedance

We write the admittance of X, in series with R as

$$Y = \frac{1}{R_L + jX_1} \times \frac{R_L - jX_1}{R_L - jX_1} = \frac{R_L}{R_L^2 + X_1^2} - \frac{jX_1}{R_L^2 + X_1^2}. \qquad (2.1)$$

The magnitude of X_1 is thus chosen to bring the real part of this admittance to the design value. Then X_2 is chosen of the right magnitude *and sign* to reduce the total imaginary part of the admittance to zero. Note that to produce the right magnitude of the real part of admittance, the sign of X_1 matters not at all, because it is squared in the expression for the real part of admittance. However, whatever the sign of X_1, X_2 must be chosen of the correct magnitude and sign to cancel the imaginary part of the admittance of the series connection.

Example 2.1.2

A 20 ohm load is to be modified by L-section to present a real part of 50 ohms impedance, at 200 MHz, using a series capacitor and a shunt inductor. Find the magnitude of each reactive element.

SOLUTION   We need $\dfrac{R_L}{R_L^2 + X_1^2} = \dfrac{1}{50}$, when $R_L = 20$ ohms, so, $\dfrac{20}{400 + X_1^2} = \dfrac{1}{50}$.

Cross-multiplying, $400 + X_1^2 = 20 \times 50 = 1000$.

$$X_1^2 = 600; \; X_1 = 24.49 = \frac{1}{\omega C} = \frac{1}{4\pi C \times 10^8}.$$

Solving for C, $C = 32.5$ pF.

The magnitude of reactive admittance to be matched is

$$\frac{24.49}{(100+600)} = \frac{1}{\omega L_1} = \frac{1000}{24.49 \times 1 \times 10^8 \, \pi} = 32.5 \text{ nH.}$$

EXERCISES 2.1.1 AND 2.1.2
1. *Find the magnitudes of the matching elements if X_1 is inductive.*
ANSWER $L_1 = 19.5$ nH, $C_2 = 19.5$ pF.
2. *Find the magnitude of the elements that will match a 50 ohm load to 75 ohms, at a frequency of 500 MHz.*

ANSWER   11.25 nH in series and 3.00 pF in parallel or 9.00 pF in series and 33.76 nH in parallel with the series combination.

2.1.3   Matching a Real Load that has Greater Resistance than Source

With a knowledge of the behavior of duals, we might expect that an L-section that first shunts the load impedance will reduce the effective real part of impedance, and fortunately, it does work that way. The impedance of the parallel combination is

$$\frac{jX_1 R_L}{R_L + jX_1} \times \frac{R_L - jX_1}{R_L + jX_1} = \frac{X_1^2 R_L + jX_1 R_L^2}{R_L^2 + X_1^2} = \frac{R_L}{1 + \left(\dfrac{R_L}{X_1}\right)^2} + \frac{jX_1}{1 + \left(\dfrac{X_1}{R_L}\right)^2}. \quad (2.2)$$

Thus, we shunt the load with reactance until the real part of impedance is correct, after which we add the series reactance of the right sign to reduce total reactance to zero.

Example 2.1.3
Find the value of shunt capacitance and series inductance to match 80 ohms real to a 50 ohm source at 500 MHz.

SOLUTION   We need

$$\frac{80}{1 - \left(\dfrac{80}{X_1}\right)^2} = 50; \ 50\left(\frac{80}{X_1}\right)^2 = 30; \ X_1 = 80\sqrt{\frac{5}{3}} = 103.3;$$

$$\frac{103.3}{1 + \dfrac{5}{3}} = 10^9 \, \pi L; \ L = 12.33 \text{ nH.}$$

1. *Find the shunt inductance and series capacitance which will make the match of impedance needed in the example above.*

ANSWERS   32.87 nH, 8.22pF

2. *Design two possible L-sections to match a load of 100 to 75 ohms.*

ANSWERS   55.1 nH shunt and 7.35 pF series; or 1.84 pF shunt and 13.78 nH series.

2.2   Use of the Smith Chart in Lumped Impedance Matching

When the Smith chart is augmented to show lines of both constant normalized impedance and admittance simultaneously, it can be very helpful in understanding the mechanics of lumped circuit impedance matching. In general, one should normalize all impedances to the generator impedance to which it is wished to match, if that impedance is purely real. Consider the examples in the previous section. Normalizing the 20 ohm load to 50 ohms

gives $r = \left(\dfrac{20}{50}\right) = 0.4$; so we enter the Smith chart at point A on Figure 2.2.

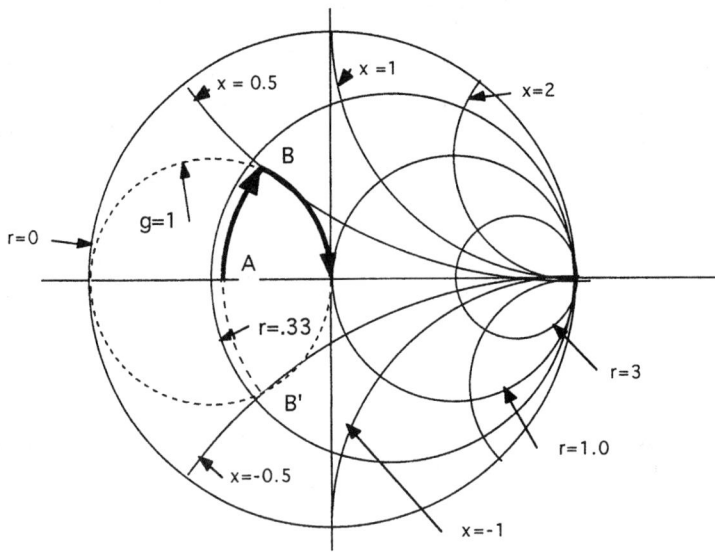

FIGURE 2.2
Transforming real impedance upwards using series reactance.

We may move in *either direction* along the r = 0.4 line until we reach the g = 1.0 line at either B or B'. From the chart, it appears that we had to add a normalized reactance of precisely 0.5. It is almost always likely to be preferable to make the more precise numerical calculation, which recommends adding a normalized reactance of $x = +0.49$. Since the frequency is 200 MHz, the series element is either

$$L = \frac{0.49 \times 50}{4\pi \times 10^8} = 19.5 \; \mu H, \; or$$

$$C = \frac{1}{4\pi \times 10^8 \times 0.49 \times 50} = 32.48 \text{ pF.}$$

At the location where $z = 0.4 \pm j0.49$, we now may calculate the normalized admittance as $y = 1 \pm j1.225$, a number which would be borne out by careful scrutiny of the two-color Smith chart. Thus, if our shunt element were an inductor, the match is completed using a series capacitor given by

$$1.225 = 50\omega C; \; C = \frac{1.225}{50 \times 4\pi \times 10^8} = 19.5 \text{ pF.}$$

If the shunt element was a capacitor, the impedance match is completed using a series inductor figured as follows

$$1.225 = \frac{50}{\omega L}; \; L = \frac{50}{1.225 \times 4\pi \times 10^8} = 32.8 \text{ nH.}$$

Next, let us consider the case where the load to be matched is 80 ohms; we enter the Smith chart, noting as seen in Figure 2.3,

$$\text{where } z = 1.6; \text{ hence, } y = g = 0.625.$$

We may move either upward or downward along the 0.625 line, although because it is not drawn on the graph, it appears that the imaginary part of the admittance, called susceptance and given the symbol "b," is approximately ± 0.5 at the point, corresponding to 100 ohms. Of course, the more accurate numerical calculation would yield 103.3 ohms. As we saw before, to obtain the impedance of 103.3 ohms, we required 3.08 pF or 32.87 nH at 500 MHz.

Again, from the Smith chart, we may read the normalized impedance after we add the shunt element as $z = 1 + j0.75$; this would mean we must add ±37.5 ohms reactive to cancel the reactance we have, but again, the more accurate numerical calculation would call out 38.7 ohms. Clearly, the accuracy of the graphical calculation is limited, but in view of the available magnitudes of the components one can buy, the graphical accuracy may be sufficient.

Perhaps the most important advantage of using the Smith chart on these impedance matching problems is that it helps the engineer to visualize a surprising number of additional possibilities, which might not otherwise have

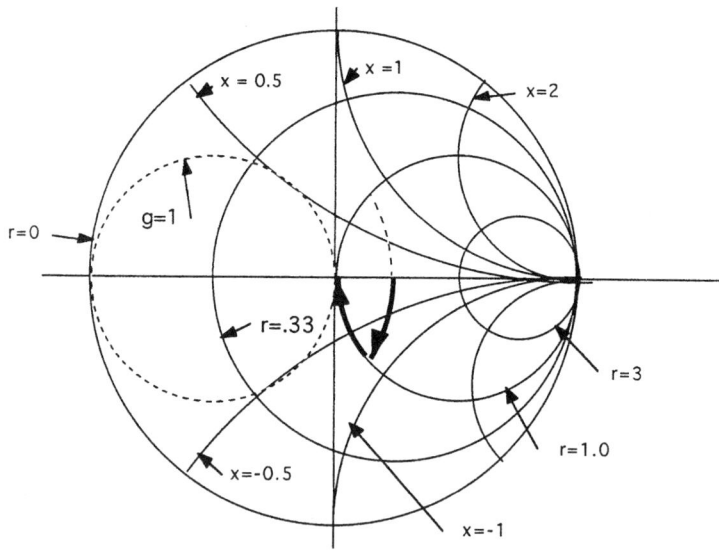

FIGURE 2.3
Transforming real impedance downward using shunt reactance.

occurred to him/her. First, we should note that z = 20 ± j anything may in principle be matched by adding a series inductor or capacitor of the correct amount to make the normalized reactance = ± 0.49. Having done that, we put in parallel the same inductor or capacitor that was needed in the original design. In examples where we start with some small load resistance, practical considerations may suggest that the series reactance one adds should be the choice which requires the smaller amount of reactance, i.e., if the load reactance is small and inductive, one should add more inductance to reach the desired impedance.

When the normalized admittance or impedance to be matched lies outside both the r = 1 and g = 1 circles, an interesting new possibility is found; we find that the number of matching possibilities doubles once again. To stay on somewhat familiar ground, let us consider for the moment impedances with the real part of 20 ohms. The added possibilities occur when x > 0.49 (of course, when x > 0.49, we may equally well start with a shunt resistance). Let us consider for example a load with normalized impedance of z = r + jx = 0.4 –j0.8. (Please refer to Figure 2.4.)

Certainly, we could again add x = 0.31 to arrive at 0.4 –j0.49 as before. Then, also as before, the normalized admittance is y = 1.0 +j1.225, so that adding more inductance to give an admittance of –j1.225, matches the line. However, we could also start by first changing admittance to get the real part of normalized *impedance* to unity. The original normalized admittance is 0.5 + j1.0. We could add normalized admittance of –j0.5, arriving at y = 0.5(1 + j), which corresponds to normalized impedance of 1 – j. Again we must add inductance,

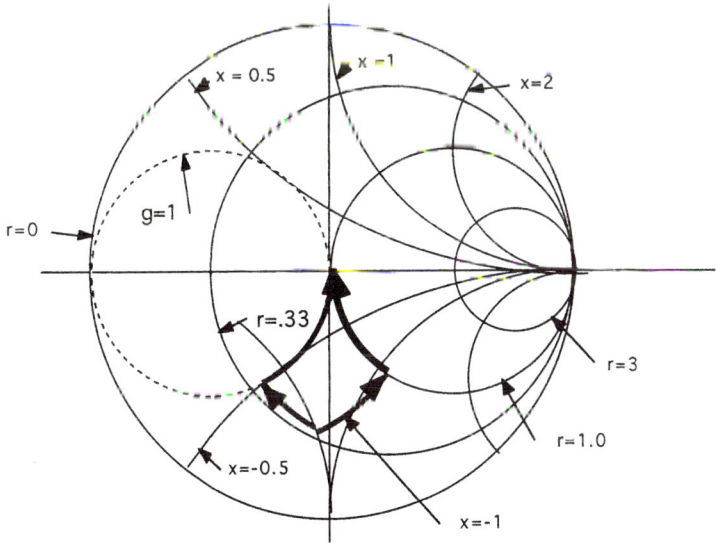

FIGURE 2.4
Illustrating two trajectories by which z – 0.4 – j0.8 may be matched.

this time in *series*, to cancel the negative reactance. The magnitude of these elements at 500 MHz is

$$\text{shunt L} \quad 0.5 = \frac{Y}{Y_o} = \frac{Z_o}{\omega L};$$

$$L = \frac{Z_o}{0.5\omega} = \frac{50}{0.5 \times 10^9 \pi} = 31.8 \text{ nH};$$

$$\text{series L} \quad 1.0 = \frac{\omega L}{Z_o};$$

$$L = \frac{Z_o}{\omega} = 15.9 \text{ μH.}$$

Theoretically, of course, one could add considerably more shunt inductance to arrive at the point

$$y = 0.5 - j0.5, \text{ and } z = 1 + j1.$$

However, the more drastic change in the circuit is probably much less practical; the larger the change in the circuit for the purpose of impedance matching, the faster the situation deteriorates as frequency changes. Of course, in this latest solution, with this shunt element the matching would be completed by adding a series *capacitor*.

EXERCISES 2.2

1. *Find the L-section elements to match a load of 10 + j30 ohms to a 50 ohm source in two ways. Frequency is 800 MHz.*

ANSWER   Series $C = 19.89$ pF, shunt $C = 7.96$ pF or shunt capacitance of 3.98 pF followed by series C of 3.98 pF.

2. *Find two L-sections which will match a load of 15 + j45 ohms to a 50 ohm line at 800 MHz.*

ANSWER   9.04 pF in series with 6.09 pF parallel with the combination; or 2.1 pF in parallel, followed by 2.8 pF in series with the combination.

2.3   Impedance Matching with a Single Reactive Element

Because impedance varies as one moves along a transmission line, one can match the line using a single reactive element if one finds the correct location. One simply moves along the line to a point where the real part of impedance is possible to work with series matching elements.

Let us illustrate what is needed by use of an example. Let us see what is required to match a 100 ohm resistor to a 50 ohm line, at 500 MHz. It is our plan to find the location nearest the load where adding a single parallel reactance will match the line; therefore, we first find the nearest location to the load terminals where normalized admittance $y = 1.0 \pm jb$. We first normalize the load impedance and enter the Smith chart at that point. Normalized load impedance is

$$z_r = \frac{Z_t}{Z_0} = \frac{100}{50} = 2.0.$$

We use the Smith chart in Figure 2.5 to graph the variation of normalized impedance as one moves from the load terminals toward the generator. $z = 2.0$ is on the positive real axis. Making the usually good assumption that attenuation effects can be neglected in a line a fraction of a wavelength long, we rotate clockwise on the circle $|\Gamma(d)| = 1/3$ until we first intersect the $g = 1.0$ line at approximately $y = 1 + j0.7$. The line will be matched from the latter location to the generator if we shunt the line with $y = -j0.7$ at this location, clearly an inductor, given by

$$0.7 = \frac{Y}{Y_0} = \frac{Z_0}{\omega L}; \; L = \frac{50}{0.7 \times 10^9 \pi} = 22.7 \text{ nH}.$$

If we work as carefully as possible with a full-sized Smith chart, we find that we moved from 0.250 to 0.417 wavelengths toward the generator, or 0.167 λ.

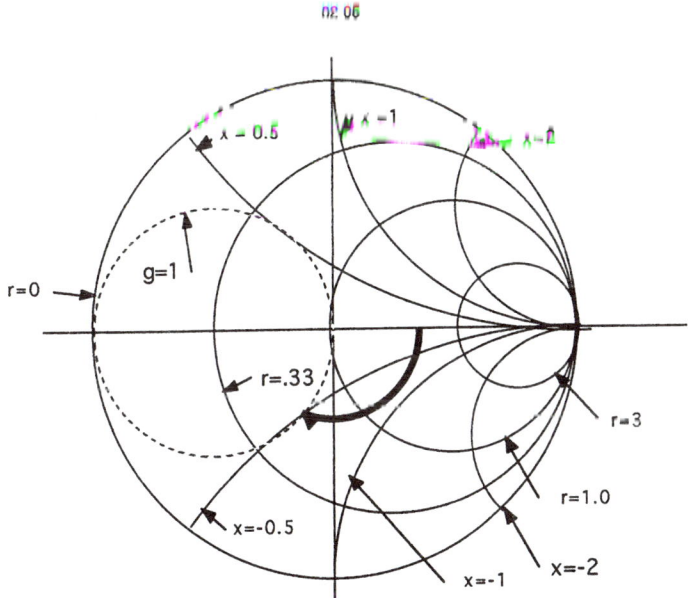

FIGURE 2.5
Moving on Smith chart from r = 2.0 to g = 1 + jb.

(A careful computation on a $100 calculator gives b = 0.7071 and a distance from load to matching element of exactly one-sixth of a wavelength. See Appendix A for some guidance in this calculation.) We enter the chart on the negative real axis at g = 0.5. We rotate on a circle of constant radius, towards the generator, until we encounter the circle where y = 1 ± 0.71; this requires moving a distance of 0.152λ. At that location, the line is matched if we add a shunt reactance having a normalized admittance of –0.71. Calculating the amount of inductance required,

$$\frac{Y}{Y_o} = YZ_o = 0.71 = \frac{50}{\omega L}; \ L = \frac{50}{0.71\omega} = \frac{50}{0.71 \times 10^9 \pi} = 22.4 \ \text{nH}.$$

Knowing the characteristics of the transmission line we are matching, we can figure the location of the matching reactance in centimeters, inches, or whatever is appropriate. Almost all coaxial lines that one could encounter are insulated with "stabilized polyethylene," for which the dielectric constant is 2.26. Hence, the velocity of waves on the line at high frequency is given as

$$\frac{c}{\sqrt{2.26}} = 1.996 \times 10^8 \ \frac{\text{m}}{\text{s}}.$$

Thus, the distance to the matching reactance, previously found on the Smith chart to be 0.152λ, can now be calculated as

$$d_{match} = .152 \times \frac{1.996 \times 10^8}{5 \times 10^8} = 6.1 \text{ cm.}$$

If the transmission line is microstrip, one must determine the effective dielectric constant, as was demonstrated in Chapter 1; one then computes wavelength using this effective value.

EXERCISES 2.3
1. *Find the location and magnitude of a matching reactive element to match 150 ohms to a 50 ohm coaxial line at 600 MHz.*

ANSWER   3.78 cm from load, 11.5 nH.
2. *Find the location and magnitude of matching reactance to match 15 ohms to 75 ohm microstrip line for which* $\varepsilon_r = 2.0$. *The frequency is 750 MHz.*

ANSWER   1.87 cm from load, 5.0 pF.
3. *Find the location and magnitude of a reactive element to match* $50 - j100$ *ohms to a 50 ohm line. Frequency is 600 MHz and the line is a microstrip for which the effective dielectric constant is 3.5.*

ANSWER   6.63 nH located 3.37 cm from the load
4. *Find the location and magnitude of the reactive element to match* $25 + j25$ *ohms to 50 ohm coaxial line. The frequency is 750 MHz.*

ANSWER   4.2 pF located right at the load.

2.4   Stub Matching

At one time, there did not exist reactive elements suitable for matching microwave transmission lines. The technique used was the called "stub matching." The required reactance was provided by a short length of transmission line having a short or open circuit as termination. With a reflection coefficient having a magnitude of unity, the input impedance of this short length of transmission line, which is what is called a "stub," is purely reactive. Hence, the appropriate magnitude of reactance is made by appropriate choice of its length.

In principle, either a shorted or open-circuited transmission line would serve equally well as an impedance matching section; in practice it may be much easier to fabricate a good solid short than an open. Too often, a signal bearing conductor which is not grounded may act as a fairly efficient antenna. If energy is lost by radiation, there will be seen a "radiation resistance," which means that the input impedance will not be purely imaginary.

The technique of stub matching may be illustrated using the same example as we treated in the previous section. There we found that in matching

100 ohms to 50 ohm line, we first found the location 0.152λ from the load. At that point, normalized input impedance was $y = 1 \pm j0.71$. The stub must thus provide $b = 0.71$

Let us return to theory to consider the input impedance of a short-circuited lossless transmission line. For a short-circuit termination,

$$\Gamma_t = \frac{0 - Z_o}{0 + Z_o} = -1; \quad \text{then,}$$

$$Z(d) = Z_o \frac{1 + \Gamma_t e^{-j2\beta d}}{1 - \Gamma_t e^{-j2\beta d}} = Z_o \frac{e^{j\beta d} - e^{-j\beta d}}{e^{j\beta d} + e^{-j\beta d}}.$$

The sum and difference of positive and negative exponentials are conveniently expressed in terms of trigonometric functions:

$$Z(d) = Z_o \frac{j2\sin\beta d}{2\cos\beta d} = jZ_o \tan\beta d.$$

Because we usually require some normalized admittance, we write

$$y(d) = \frac{1}{\dfrac{Z(d)}{Z_o}} = \frac{1}{j\tan\beta d} = \frac{-j}{\tan\beta d}.$$

In the example, we require, $\dfrac{-j}{\tan\beta d} = -j0.71$;

$$\therefore \beta d = \frac{2\pi d}{\lambda} = \tan^{-1}\frac{1}{0.71} = \tan^{-1}1.42 = 0.957 \text{ rad}$$

$$d = \frac{0.957\lambda}{2\pi} = 0.152\lambda$$

The reader may be very impressed by the fact that the stub length and distance from the load came out to be the same in this example. *This was sheer coincidence* and will not happen again. Young engineers should not expect to see any appreciable amount of real magic in their careers.

EXERCISES 2.4

1. *Find the length of stubs made with 50 ohm insulated line with polyethylene which would replace the discrete reactances in the exercises of the previous section*

ANSWERS   Stub susceptance of $b = 1.1547$ is provided by a length of 3.78 cm.

Stub susceptance of b = 1.7889 is provided by a length of 11.85 cm.
Stub susceptance of b = –2 is provided by a length of 2.46 cm.
Stub susceptance of b = 1 is provided by a length of 9.98 cm.

2.5   Quarter-Wavelength Matching Sections

There is one more traditional method of matching loads to transmission lines or to generators, which in its simplest form, matched a real impedance to another real impedance. Hence, it might be considered not versatile enough for general use. One example of its success is shown in Figure 2.6.

02-06

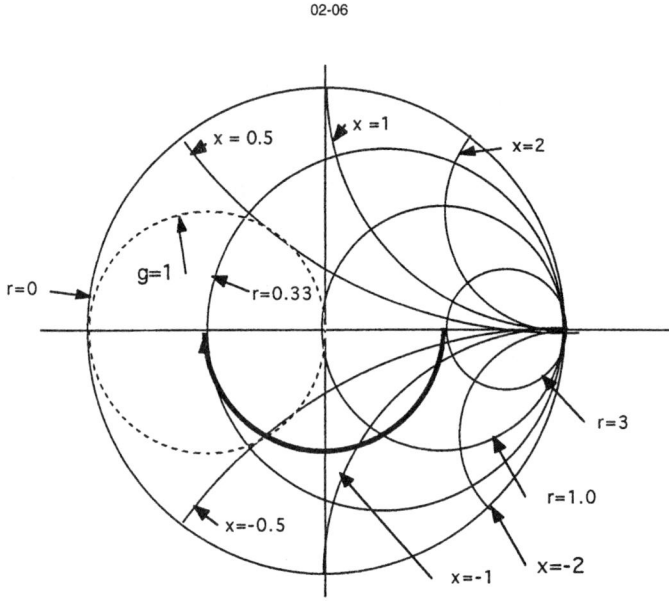

FIGURE 2.6
Quarter-wavelength line transforming r = 3 to r = 1/3.

On this Smith chart, we see an arc rotating one-half of the way around the chart, which has the effect of transforming a normalized impedance of 3.0 to one of 1/3.0. It can be seen that the necessary design choice is to choose the characteristic impedance of the quarter-wavelength section appropriately. It is not difficult to prove that given two real impedances which one wishes to match to each other, the characteristic impedance of the quarter-wavelength section needs to be the geometric mean of the two impedances; one example of how this could work would be for a load impedance of 225 ohms and a generator impedance of 25 ohms. The characteristic impedance of the matching section would then need to be

$$Z_n = \sqrt{Z_1 Z_2} = \sqrt{225 \times 25} = 75 \text{ ohms.}$$

EXERCISE 2.5.1

1. *Find the impedance of the quarter-wavelength matching sections to match a load of 150 ohms to a generator impedance of 50 ohms, a 200 ohm load to a 50 ohm generator, and a 750 ohm load to a 120 ohm generator.*

ANSWERS   86.6 ohms, 100 ohms, and 300 ohms.

2.5.1   Matching with Other than Quarter-Wavelength

For many years this writer thought that he and his students were the only people in the world who had noticed that a complex impedance can be matched to a real one by using matching sections of more or less than one quarter-wavelength. Finally, there emerged one author named Rizzi[*] who succinctly specified that a complex load impedance R + jX can be matched to a generator or transmission line having source or characteristic impedance Z_0 using a "matching section" having characteristic impedance

$$Z_{01} = \sqrt{\frac{Z_0\left(R(Z_0 - R) - X^2\right)}{Z_0 - R}}.$$

Rizzi also gives the length of the matching section as "l" given by

$$\tan\beta 1 = Z_{01}\frac{Z_0 - R}{XZ_0}.$$

There are one or two possible complications which may arise in the use of these formulas. In the characteristic impedance formula, the minus signs present may lead to a maximum value of reactance that can be matched; greater values of reactance lead to a purely imaginary characteristic impedance, which is of course not available. In the formula for length of the matching section, first it should be insisted that the dimensions of βl *must* be radians. Another unfortunate characteristic of both calculators and most computer programs is that if either is asked to find arctan of something negative, it will give an answer in the fourth quadrant as a negative angle. Such an answer would imply a negative length of matching section, which makes no physical sense. The reader is urged to obtain an answer in the second quadrant in such cases; the simplest way may be to add pi radians to the answer in the fourth quadrant.

Example 2.5.1

Let us check out this method. A rather bulky antenna called the "rhombic" has an input impedance of 800 + j175. Find the characteristic impedance and

---

[*] Peter A. Rizzi *Microwave Engineering, Passive Circuits,* Prentice Hall, Englewood Cliffs, NJ, 1988, pp. 132–133.

length of matching section to match it to 300 Ω. (*Note*: 300 Ω is the characteristic impedance of the balanced two-wire line called "twinlead.")

SOLUTION   Substituting in the expression above,

$$Z_o' = \sqrt{\frac{300\left(800^2 - (300 \times 800) + 175^2\right)}{800 - 300}} = 508.3 \ !$$

Next, let us find the length of the matching section in a way that capitalizes on our knowledge of the Smith chart. First, we normalize the load impedance by dividing it by the characteristic impedance **of the matching section** and plotting it on the Smith chart. We then rotate on an arc of constant reflection coefficient to the negative part of the real axis, obtaining $l = 0.283\lambda$ as the length of the matching section. See chart in Figure 2.7.

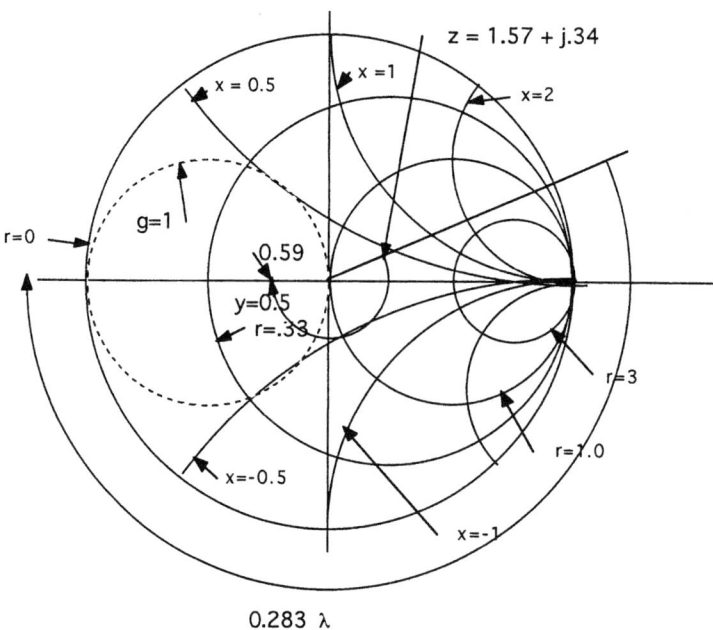

0.283 λ

FIGURE 2.7
Smith chart normalized to 508 Ω for matching load of 800 + j175 Ω to 300 Ω.

We know that we must end up on either the positive or negative real axis, where the input impedance is purely real; since the amount of the normalized impedance is 0.590, we know we want the *negative* real axis. The point we plotted is found to be 0.283 "wavelengths toward load" on the Smith chart, which then is the length of the matching section in wavelengths. Let us check this result using Rizzi's formula. We have

$$\tan \beta 1 = \frac{508.3 \times (300 \quad 800)}{175 \times 300} = -4.841.$$

For such a tangent, the HP 48S gives $\beta l = 1.3671$. Since we cannot use this answer, we add pi, obtaining $\beta l = 1.7745$. Now, we must use the relation between wavelength and phase shift constant, $\beta = 2\pi/\lambda$; so solving for the length "l" of the matching section, "l" $= 0.2824\lambda$, which agrees reasonably well with our graphical determination on the Smith chart.

EXERCISES 2.5.1
1. Match 200 − j50 to a 300 ohm line.

ANSWER $Z_{01} = 230\ \Omega$; length of 0.342λ).
2. Find the characteristic impedance and length of a matching section to match a load impedance of 35 + j7 to a 50 Ω line. (Note that while load resistance is less than the impedance to be matched to, the reactance is not large enough here to make the match impossible in principle.)

ANSWER $39.8\ \Omega$, $0.166\ \lambda$
3. Repeat step 1 for a load of 100 + j70.

ANSWER $150.5\ \Omega$; $0.152\ \lambda$.

2.6 When and How to Unmatch Matched Lines

In Chapter 1, we found that load and source impedances needed to be far from 50 ohms for some transistors to be conjugate matched at both the input and output ports. That makes it sound a little as though one may be in the business of changing a nice real 50 ohms into something complex, and one may wonder if he/she knows how to do that.

As usual on transmission lines, we may obtain Z(d) as

$$Z(d) = Z_0 \frac{1 + \Gamma(d)}{1 - \Gamma(d)}.$$

Hence, if one acts up for simultaneous impedance match at the input and the output, the source impedance "seen" by the transistor will be

$$Z_s = Z_0 \frac{1 + \Gamma_{MG}}{1 - \Gamma_{MG}} = 50 \frac{1 + 0.890\angle -178.71^\circ}{1 - 0.890\angle -178.71^\circ} = 2.91 \quad j0.561.$$

Of course, the implication of this result is that the input impedance of the transistor will be 2.91 + j0.561 if Γ_{MT} is connected as the load. Hence, the design goal of the engineer is to modify a real impedance of 50 ohms so as to obtain $2.91 - j0.561$. His/her procedure is to observe first that the real part of impedance must be greatly reduced. If the engineer decides to accomplish this using an L-section, the first step is to connect a reactance across the 50 ohm input impedance. As was found in Equation 2.2 in the earlier part of this chapter, the amount of the reactance is given by

$$2.91 = \frac{50}{1 + \left(50/X_1\right)^2}.$$

Solving this equation for X_1, one then obtains the *magnitude* of the reactance as

$$\left|X_1\right| = 12.43 \text{ ohms.}$$

We have not absolutely determined the *sign* of the reactance. The other relation determined in Equation 2.2 at the beginning of the chapter was

$$X_2 = \frac{X_1}{1 + \left(X_1/50\right)^2}.$$

The first thing to be noted here is that the sign of X_2 will be the same as X_1. From this circuit, we need a final reactance that is negative, so the first principle here is that every time we make a choice, it should get as close as possible to the final result. Certainly here, we could use a positive reactance and cancel it out with a much larger negative reactance, but better engineering practice would be to start with the correct sign and simply add to it. From the equation just above, if we start with $-j12.43$ in parallel with 50 real ohms, the result is $X_2 = -11.71$. But we needed only a total of -0.56, so we must add $+j11.15$ ohms in series to get the correct result.

Now, we have proven that the transistor can be provided the impedance it desires. It might be comforting to prove that the generator, which has a purely real impedance of 50 ohms, "sees" the input impedance it expects. It may be helpful to refer to Figure 2.8.

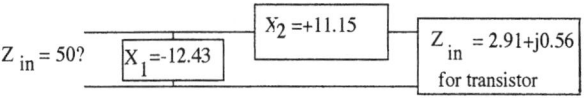

FIGURE 2.8
Matching L-section at the input.

This is to reinforce our statements of what is connected where:

- We have connected j11.15 in series with 2.91 + j0.56, obtaining a total of 2.91 + j11.71.
- Now we wish to know the *admittance* of this series combination, so we take the reciprocal of this series impedance, obtaining $Y = 0.01999 - j0.0804$.

This is in parallel with a reactance $Z = -j12.43$, which corresponds to an admittance of j0.08045, so the reactances cancel each other and we are left with a real impedance which corresponds very closely to 50 ohms. (It should be noted that the small discrepancies one finds here result from pulling numbers off the calculator and resubstituting them with limited accuracy.)

EXERCISES 2.6

1. Repeat the exercise above at the output of the transistor, where one is looking for $\Gamma_{ML} = 0.777\angle 66.10°$.

ANSWERS   One needs $Z_L = 20.34 + j72.92$ ohms. Because this real part is smaller than 50 ohms, one first needs to parallel the 50 ohms with the appropriate reactance; from Equation 2.2, this would be $|X_1| = 41.41$.

The series equivalent of this parallel combination is 20.34 + j24.56, if the sign of X_1 was chosen positive.

Now, one still must add series reactance to arrive at the right total; $X_2 = 48.36$ would be the right amount to add.

2. The values of Γ_{ML} and Γ_{MG} used above were found for the MRF 571 transistor at 1 GHz. Use this knowledge to find the kind and magnitudes of lumped circuit elements to perform the conjugate impedance matching.

ANSWERS   At the input, parallel the generator with capacitance of 12.8 pF. Then place an inductor of 1.77 nH in series with the parallel combination.

At the output, parallel the 50 ohm load with j41.41, which will require an inductor of 6.59 nH. In series with this, we need reactance of j48.36, which calls for an inductor of 7.7 nH.

3-A. Find the length and characteristic impedance of matching sections to produce the correct driving and load impedances, i.e., $Z_G = 2.91 - j0.56$ and $Z_L = 20.34 + j72.92$ ohms.

ANSWERS   At input, $Z_{01} = 12.08 \Omega$, length of 0.2578λ.

At output, matching section method won't work because reactance is too high.

3-B. Assume the matching section is to be made using microstrip material for which the dielectric constant is 2.25. If the thickness of the dielectric is one millimeter, find the width of the trace for Z_{01} found in A, the effective dielectric constant, and the length of the matching section at $f = 1$ GHz.

ANSWERS   W = 1.82 cm, ε_{eff} – 2.11, Length   5.02 cm.

4. *Find the distance from the input or the load terminals of the above transistor where a single series reactance may be placed in a 50 ohm line to conjugate-match input and output lines. Also, find the kind and the magnitude of the reactance. Assume that the line is microstrip as analyzed in Section 1.7, having $\varepsilon_{eff} = 1.905$.*

ANSWERS 0.2105λ or 4.57 cm for input terminals.

     Need $x = -3.904$, or $-j195.2$ Ω, which could be provided by 0.815 pF. Go 0.354λ from load terminals, or 7.69 cm, add $x = -2.47$, or $-j123.4\Omega$, which can be provided by 1.29 pF.

5. *Find the distance from the input or the load terminals of the above transistor where a single short-circuited shunt stub in 50 ohm line may be placed in the 50 ohm line to conjugate-match input and output lines. Also, find the physical length of the stub.*

ANSWERS   In input line, go 0.0359λ or 0.78 cm from input terminals. For normalized input admittance of j3.904, need 0.230λ, or 5.00 cm. In output line, go 0.104λ, or 2.26 cm. For normalized stub admittance of $y = -j2.47$, use stub length 0.0306λ, or 0.67 cm.

2.7   Designing Reactances for Waveguides

The next topic may seem more like a digression than a pertinent topic. Its rationale is that in an academic department where availability of modern equipment is limited, but there does remain some old waveguide hardware, it provides the procedures for what can be a surprisingly successful lab experiment. In an area where it can be quite rare to obtain results that represent the students' expectations at all, this subject can yield results that are more accurate for the most careful experimental technique. Here, we will not develop the theory needed for waveguide communications, but will simply quote the pertinent aspects of theory.

     All of the standard waveguides which were developed are rectangular pipes having outside dimensions, one of which is exactly twice the other. An infinite number of modes can be transmitted through any waveguide, where any mode is a set of possible electric and magnetic fields. The waveguide is a high-pass filter for each mode, and waveguides are almost always operated at such a frequency of operation that the frequency is above the cut-off frequency for only one mode, named the dominant mode. The standard cross-section of waveguide is as shown in Figure 2.9.

     Cut-off frequency for the dominant mode is given as $f_c = c/(2a)$. The most common waveguide found in undercapitalized academic labs may be designated WR 90, where the 90 gives the larger of the inner dimensions "a" in $(1/100)$ths of an inch. If the reader is slightly startled to find English units being standard in waveguide dimensions, one needs to recognize that the United States and the United Kingdom developed radar during World War II;

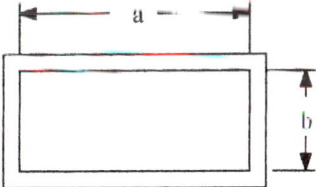

FIGURE 2.9
Waveguide geometry and dimensions.

so, at least in these nations, inches still hold sway. It is recommended that the engineer take the inches he/she is given or obtains and converts them to meters or centimeters at the earliest opportunity. It is certainly true that one will not go far wrong in using "c" $= 3 \times 10^8$ meters/second, whereas the speed of light as 186,000 miles/second that we of the "silent generation" learned was never a very valuable constant to remember.

Both signs of reactance are available by suitable design. However, there are practical aspects which militate against capacitive reactances. The two types serve to reduce one of the dimensions: inductive forms reduce the larger of the dimensions, whereas capacitive ones shorten the shorter dimension. In the dominant mode, electric field lines point in this shorter direction. Now, of course, the main reason for using a waveguide at all, rather than some kind of cable, to transmit power, is that a waveguide permits transmission of much more power, compared, say, to coaxial cable, before arcing, which is a high voltage breakdown of the dielectric. However, if one puts in a structure that tends to shorten the more critical dimension of the waveguide, it is most likely that one has "shot oneself in the foot." This is probably the main reason why capacitive reactance is fairly seldom designed in waveguides.

The structures to be designed here apparently reminded the early experimenters of the iris of the eye, which is of course the part that surrounds the opening in the eye, the pupil. Since the word is also used on the graphs, we will not struggle against it, but will also call them irises. As can be seen in Figure 2.11, one may use a symmetrical or an asymmetrical geometry for these irises. This author's recommendation is that the engineer use the simplest form, as is the only one shown in Figure 2.10.

Having said all these things, let us design at least one inductive iris and a capacitive one, using these curves. Because in English-speaking lands we are given our dimensions in inches and, besides, there is no hope of American machine shops working in anything but inches anytime soon, we will also output our answers in the form of inches.

Example 2.7 Inductive Iris

Design an inductive iris to have a normalized admittance of 1.5, at a frequency of 9.0 MHz.

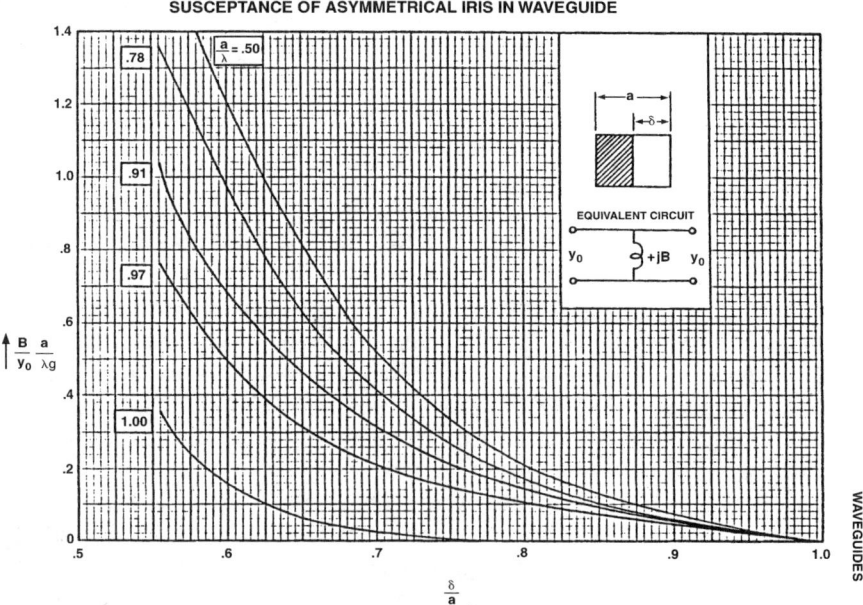

FIGURE 2.10
Inductive irises.

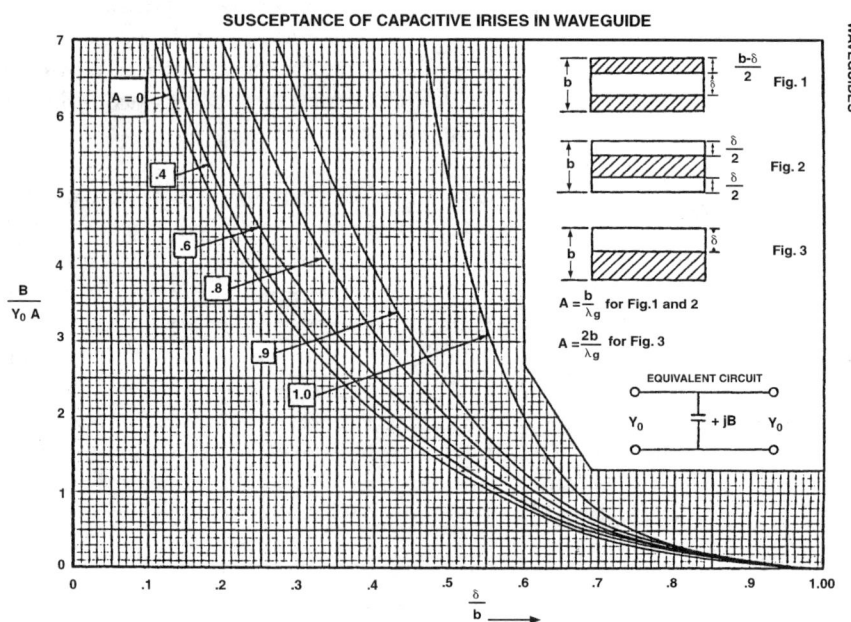

FIGURE 2.11
Capacitive irises.

SOLUTION   The beginner may be a bit comprised by the form of some param eters used in iris design, as well as in the labeling of the vertical and horizontal axes.

Looking first at the vertical axis on Figure 2.10, we see the labeling is $\dfrac{b}{Y_o} \times \dfrac{a}{\lambda_g}$. First, our problem statement has called out the value of B/Y_{ou} as 1.5. WR 90 waveguide has a = 0.9 *inches*. So we next must find the wavelength inside this waveguide at an operating frequency of 9.0 GHz. First, the free space or plane wave wavelength would be

$$\lambda = c/f = 3 \times 10^{10} \text{ cm/s} / 9 \times 10^9 \text{ Hz} = 3.3333 \text{ cm}.$$

Dominant mode cutoff frequency for this waveguide is

$$c/2a = 3 \times 10^{10} \text{ cm/s} / 2 \times 0.9 \text{ inch} \times 2.54 \text{ cm/inch} = 6.56 \times 10^9 \text{ Hz}.$$

Therefore,

$$\lambda_g = \frac{\lambda}{\sqrt{1 - \dfrac{[f_c]^2}{[f]}}} = \frac{3.333}{\sqrt{1 - (6.56\ 9)^2}} = 4.870 \text{ cm}.$$

Now, finally, we may enter the graph at $1.5 \times 0.9 \times 2.54/4.87 = 0.704$. Oops, we have not yet found the parameter $a/\lambda = 0.9 \times 2.54/3.333 = 0.686$. Suppose, we make a linear interpolation between the value 0.78 and 0.5; we can say 0.686 is roughly 1/3 of the distance from 0.78 to 0.5. There are almost exactly 6 of the little squares to be seen. So, we read off the quantity $\delta/a = 0.650$ almost exactly. That ratio is the fraction of the waveguide *not* blocked off by the iris. So, now we want our machinist to make a cut into the waveguide that is 0.9 inches x 0.35, or 0.315 inch. To this, we must add the wall thickness for WR 90, which is 0.050 inches, or a total of 0.365 inches. He makes the width of the cut appropriate for whatever thickness of sheet copper there is available, which perhaps another technician will silver solder into place. It then goes to the so-called "plating shop," where the technician cleans it all up beautifully using some of the most toxic liquids known to man. It may then be tested to see how well we designed it.

Example 2.8 Capacitive iris

It is certainly conceivable that one will sometimes use waveguides at power levels well below the maximum possible. In such a case, there is no compel ling reason not to use a capacitive iris; so, let us design for $B/Y_{ou} = 1.5$, at a frequency again of 9 GHz.

SOLUTION   Now, the curves given in Figure 2.11 give three different geometries for capacitive irises, but our previous advice to keep the design as simple as possible points immediately to what is called "Fig. 3" on that curve. We have not changed the waveguide, frequency, or the fact that we use the dominant mode and, so, cut-off frequency and, hence, guide wavelength are the same as in the other example. Thus, $\lambda_g = 4.870$ cm.

There is only one parameter to compute, A = (2b)/ λ_g = (2 × 0.4 inches × 2.54 cm/inch)/ 4.870 cm = 0.417. Now, we must enter Figure 2.11 on the vertical axis at B/Y_o A = 1.5/0.417 = 3.60.

Next, the curves for A = 0.40 and 0.60 are very close together, so we scan across the graph for B/Y_o A = 3.60 to A = 0.4, and read the fraction of the waveguide left open as $\delta/b = 0.270$. Hence, we direct our machinist to saw into the waveguide that amount plus a wall thickness (which is also 0.050 inches on that side), or Cut = (1 − 0.270) × 0.40 inches + 0.050 inches = 0.342 inches.

EXERCISES 2.7
1. In WR 90 waveguide, at an operating frequency of 10 GHz, design asymmetrical irises (as we did above) each to have normalized reactance B/Y_o = 1.0.

PARTIAL ANSWERS   for Checking:

$$\lambda = 3.0 \text{ cm}; \ \lambda_g = 3.0 \Big/ \sqrt{1 - (6.56/10)^2} = 3.9848 \text{ cm}$$

$$a/\lambda = 0.339, \ \frac{B}{Y_o} \times \frac{a}{\lambda_g} = 0.256$$

Using the curve for $a/\lambda = 0.5$, one should leave 0.785 of the waveguide open, so the inductive iris cut should be 0.540 inches. For capacitive iris calculation, A = 2b/λ_g = 0.510. Interpolating midway between A = 0.4 and 0.6, we find 0.445 of the waveguide should be open; our cut is 0.228 in.

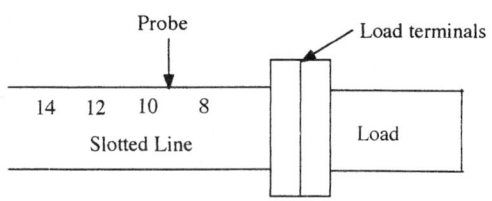

FIGURE 2.12
Detail of slotted line near load terminals.

2.8   Location of Iris from Standing Wave Measurements

Refer to Figure 2.12, on which the numbers on the slotted line increase as one moves away from the load terminals; they may thus be considered to register "d," which we have identified as the distance of the point from load terminals. The first part of any standing wave measurement might be to put a good short-circuit on the load terminals and determine the location of minima on the slotted line. Let us say they came out at 8, 10, and 12 cm.

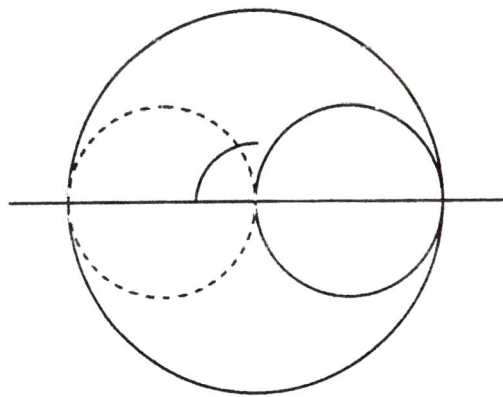

FIGURE 2.13
Smith chart for VSWR measurements.

Example 2.8

Then, suppose we connect our previously unmeasured load and find VSWR = 2.0, with minima at 9.5, 11.5, and 13.5 cm. Where should we locate an iris and what should be its susceptance?

SOLUTION   First, we know that the distance between adjacent minima on the slotted line is one *half* wavelength, thus $\lambda_\gamma = 2 \times (10-8)$ cm = 4 cm. For this VSWR, the magnitude of the reflection coefficient is $|\Gamma_t| = (2-1)/(2+1) = 1/3$. So we enter the Smith chart at $\Gamma(d) = 0.333\angle180°$ or at what is the same thing, $z = 0.5 + j0$. In terms of the markings on the Smith chart, we can say we are at 9.5, 11.5, or 13.5, whichever is convenient for us. If we need to know what is the load impedance referred to the load terminals, we must move to 8.0, 10.0 cm, or the like. From 9.5 to 10 is 0.5 cm, or $0.125\lambda_g$ *towards the generator.* When we do this, we find $z_L = 0.8 + j0.6$. Now, suppose we say we go from the load terminals to the *first location in the load structure where a shunt susceptance of either sign will match the line.* It turns out that there may be an untenable difficulty with this first location, on the g = 0 circle. We find that the input admittance at this location is y = 1 − j0.71. From the load terminals, we have gone only $0.027\lambda_g$ = 1.08 mm towards the load. However, there is a husky flange by which one bolts the load to a similar flange on the slotted section, both of which are approximately 4.5 mm thick. Clearly, if you ask the technician to saw into the flange, he will mention to your boss that he hired a real dud this time and will suspect that you were missing the day they did this calculation in your college lab. With superior wisdom, you say you will go where y = 1 + j0.71, for which you must travel 0.222λ_g, or 8.88 mm. Of course, again to avoid the scorn of the machine shop, be sure to convert those milli-meters to inches, and tell the man to make the cut 0.351 inches from the ter-minals side of the flange. Of course, from the iris, you need $y = -j0.71$, which is inductive. Now that you know how, it is left as an exercise to figure depth of cut. For convenience, assume the operating frequency is 10 GHz.

PARTIAL ANSWERS   Enter Figure 2.10 on vertical axis at 0.180, read off (doing a small interpolation) 0.805 of waveguide is open. Therefore, cut $0.195 \times 0.9 + 0.050$ inches, getting 0.226 inches.

3

Selective Circuits, Oscillators, and Receivers

Most readers have probably been exposed to selective circuits in a tuned circuits section of the ac circuits course. However, such a treatment is often rushed and rather light on practical aspects, especially high frequency phenomena. Of course, unless their company is in the business of producing selective circuits, not too many engineers will be required to do any tuned circuit design. However, it is worthwhile to introduce some of the vocabulary and ways of specifying the performance of selective circuits. Also, while the engineer needing a receiver may be able to buy integrated circuits containing electronics and fixed tuned circuits, he/she may still need to design for a particular input frequency and against specific possible interferences. Thus, this chapter will go from the performance of rather simple, single-tuned circuits to the rather high performance of piezoelectric crystal filters. In one part of the chapter, resonance will be employed as the main determinant of the operation frequency of several types of oscillators.

3.1   LRC Series Resonance

Series resonance is here analyzed, not because it is very often put to use, but because later equivalent circuits are most easily understood in terms of the series resonator. Consider the circuit in Figure 3.1

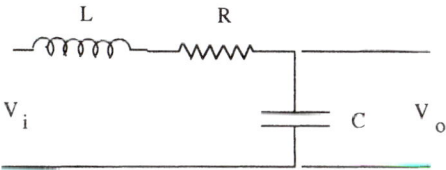

FIGURE 3.1
Series RLC resonator.

We can write the transfer function relating output voltage over input voltage by simply applying voltage division, saying that the desired ratio is the capacitive impedance over the total series impedance:

$$\frac{V_o}{V_i}(j\omega) = \frac{1/j\omega C}{j\omega L + R + 1/j\omega C} = \frac{1}{1 - \omega^2 LC + j\omega CR}.$$

Because the denominator contains a real and an imaginary part, we might suspect that the maximum of the transfer function would occur at the value of ω for which the real part goes to zero. This value, at which the inductive and capacitive impedances exactly cancel each other out, is called the frequency (in radians per second) of series resonance. We find that the transfer function can peak up rather sharply if the value of resistance is small compared to the value of either reactance.

Example 3.1.1

Let us plot the magnitude of the transfer function versus frequency for $L = 1$ H, $C = 1$ µF and for several rather different values of resistance, to wit, 20, 50, and 1000 ohms.

SOLUTION   See Figure 3.2.

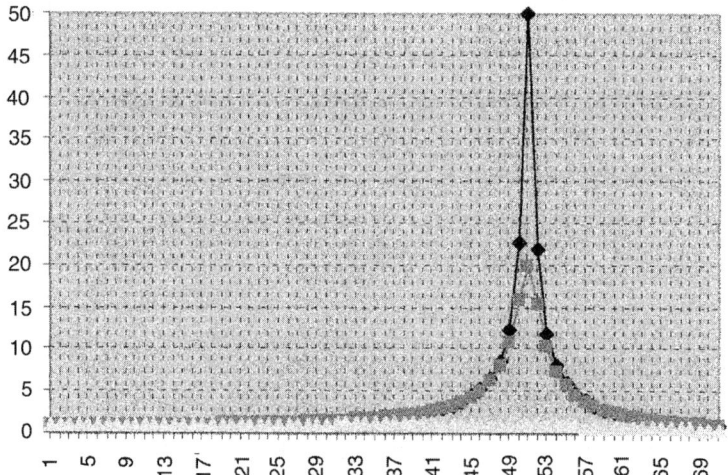

FIGURE 3.2
Series resonator capacitor voltage for several values of R.

Perhaps the most striking aspect of the graph in Figure 3.2 is that for the lowest value of R, the ratio of output to input voltage is a maximum of 50. For that or perhaps other reasons, this quantity was originally named the "Quality Factor," abbreviated Q, of the circuit. Because the source of R is in fact mainly the unavoidable resistance of the wire the inductor is wound from, the Q for series resonance is usually defined as

$$Q = \omega_0 L/R.$$

Perhaps more significantly, it is related to the half-power bandwidth of the circuit. Clearly, the transfer function drops to 0.707 of its peak value when the real part of the denominator is equal to the imaginary part. Using also the expression for resonant frequency, we may write the equation for the transfer function as

$$H(j\omega) = \frac{1}{1 - \omega^2/\omega_0^2 + j\omega/\omega_0 Q}.$$

Setting the real part equal to the imaginary part, we obtain the two positive values for half-power frequency as

$$\omega_{2,1} = \sqrt{\omega_0^2 + \left(\frac{\omega_0}{2Q}\right)^2} \pm \frac{\omega_0}{2Q}.$$

One can obtain the half-power radian bandwidth, which this author likes to call W, by subtracting the lower half-power frequency from the higher. Since both expressions contain the radical, they subtract, and we are left with

$$W = \omega_0/2Q - \left(-\omega_0/2Q\right) = \omega_0/Q.$$

This may be stated in words as, "The fractional half-power bandwidth in a resonant circuit is given by 1/Q."

Example 3.1.2

Suppose the 10 kHz bandwidth of an intermediate frequency (IF) (centered at 455 kHz) for an AM receiver is determined by one tuned circuit. What must be the Q of the tuned circuit?

SOLUTION   We can get the required Q simply by dividing the center frequency by the bandwidth. Hence,

$$Q = 455 \text{ kHz}/10 \text{ kHz} = 45.5.$$

EXERCISES 3.1
*1. Repeat the example above for RF amplifiers at each end of the AM broadcast band,
i.e., at center frequencies of 530 and 1650 kHz.*

ANSWERS   53 and 165.

It is fortunate that one will seldom have to have the latter Q, as such may be far from available.

2. Repeat the example for bandwidth of 200 kHz in FM IF amplifiers (centered at 10.7 MHz) and the high end of the band.(107.9 MHz).

ANSWERS   53.5 and 539.5.

(IF filters are often prefabricated using the high-Q capabilities of so-called ceramic filters, and one would have to be excessively innocent to attempt the latter filter.)

3. If we are aiming to pass the entire assigned 6 MHz, what Q would be required for channel 2 centered at 57 MHz, channel 13 at 213 MHz, and channel 40 at 629 MHz?

ANSWERS   9.5, 35.5, and 104.8.

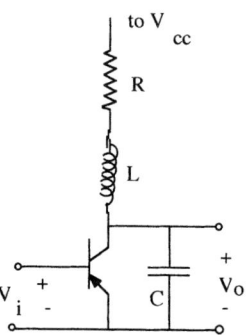

FIGURE 3.3
Parallel tuned circuit.

3.2   Parallel Resonance

Figure 3.3 shows a perhaps more commonly used selective circuit than a series resonator. The reader should remember that because power supplies generally are full of capacitors chosen to have low impedance at 60 Hz, then as far as signal is concerned, the inductor, which also has a small resistance, runs from collector to signal ground, and is in parallel with the capacitor. Then, parallel resonance occurs when the capacitive admittance balances out the imaginary part of the admittance from the series combination of R and L. Thus,

$$\omega C = \text{Im}\left(\frac{1}{R+j\omega L} \times \frac{R-j\omega L}{R-j\omega L}\right) = \frac{\omega L}{R^2 + (\omega L)^2} = \frac{1/\omega L}{1+(1/Q)^2}.$$

Here, we see that if the Q of the circuit is fairly large, the frequency of parallel resonance is almost the same as series resonance. What is dramatically different is the equivalent parallel resistance

$$\frac{R^2 + (\omega L)^2}{R} = R(1 + Q^2).$$

It is especially convenient to have a resonator of purely parallel components, because the practical resonator may also be paralleled by bias resistors. Naturally, the usual rules hold for combining parallel resistors, so the effects of shunt resistors on bandwidth are calculable. If we call the equivalent resistor R_p, the Q of a parallel resonator is

$$Q = \omega C R_p = R_p / \omega L.$$

Again, the bandwidth of a parallel resonator is given by the center frequency divided by the Q, as determined from the Q of the inductor shunted by any other resistors.

Example 3.2
An inductor of 1 µH has a Q of 40 at 455 kHz.

 a. What capacitance is required to make a parallel resonator at 455 kHz?
 b. The capacitor is shunted by actual resistors of 50 kΩ and 470 kΩ. What will be the half-power bandwidth of the combination?

SOLUTION   Let's ignore the fact that the equivalent parallel inductor is increased by the factor $1 + 1/Q^2$. Hence,

$$C = 1/\left(2\pi \times 4.55 \times 10^5\right)^2 \times 10^{-6} = 0.122 \ \mu F.$$

Until we add resistors in parallel, we can expect Q still to be 40. Then,

$$R_p = \omega L Q = 4.55 \times 10^5 \times 2\pi \times 10^{-6} \times 40 = 114.4.$$

This resistance is so small that it will not be appreciably shunted by the bias resistors. Hence, the Q will still be 40 and bandwidth will be

$$B = f_0 / Q = 455 \ \text{kHz}/40 - 11.375 \ \text{kHz}.$$

EXERCISES 3.2

1. *Now, design an FM IF amplifier. Let the inductor be 0.5 μH and have a Q of 100 at 10.7 MHz. Find C for this resonance frequency and find the amount of resistance which must shunt the tuned circuit for a bandwidth of 200 kHz at 10.7 MHz.*

ANSWERS   442 pF. $R_p = 3361\Omega$ for a Q of 100.

Since we need Q = 53.5, total parallel resistance must be 1798. We get this by paralleling 3361 by 3868.

3.3   Receiver Requirements

The signals arriving at our homes from radio and TV stations are rather low in level, typically one or two microvolts to a few millivolts. Before they can be demodulated, they may require rather a great deal of amplification. To look at what was once a rather difficult problem, in the AM broadcast band, the carrier frequencies may vary from about 540 kHz to over 1600 kHz. Severe interference of one channel with another may be encountered if one tries to amplify all channels simultaneously, and, of course, it would not be needed, because one usually only needs to receive one channel at a time. The first straw man to be knocked down in receiver circuit design is called the "tuned RF receiver." What is meant by this is that each channel would be selected by adjusting the simultaneous resonance frequency of several cascaded amplifiers. However, the width of the AM broadcast band made this impractical. An interesting facet of resonant circuits is that the Q is apt to be relatively flat, or constant, over a fairly wide range of frequency. Now, AM stations may be assigned 10 kHz apart, with examples at each end of the band at 530 kHz, 540 kHz, 550 kHz, 1600, 1610, 1620, and so on. Suppose one built a tuned RF receiver to discriminate adequately against the signals at 530 and 550 kHz when tuned to 540 kHz. This would require a circuit Q of around 540 kHz/10 kHz, or 54. However, at 1620 kHz, a Q of 54 gives a bandwidth of 30 kHz, so one would surely not discriminate well against a signal at 1600 or 1640 kHz. On the other hand, if one looked for a bandwidth of 10 kHz at 1620 kHz, the circuit Q would be 162; and while such a Q might be obtainable on rare occasions, one must not expect it very often. Even if Q = 162 were available, when such a circuit tuned 540 kHz, bandwidth would be about 3300 kHz, and this receiver would have lower fidelity than your common variety of telephone receiver.

The receiver selectivity problem was solved in an ingenious manner by the hero of many communications engineers, Edwin Armstrong. Using the proposal he might have made, he might have said, "Let's have the first step in our receiver be one in which, no matter what carrier is being received, we move it to the center frequency of a *fixed-tuned* amplifier, where most of the high frequency amplification will be done." For the AM band, this fixed-tuned circuit was placed at 455 kHz; this was below the carrier frequencies being received, but well above the frequencies of the modulation signals. Thus, this fixed-tuned amplifier was said to be at "intermediate frequency," abbreviated IF.

Now, one may ask, how is this translation of carrier frequency downwards to be accomplished? Well, it's like this. The "modulation theorem" of circuit theory says that if one achieves a multiplication of time functions, the result in the frequency domain is sum and difference frequencies. To test for this as an undesired result in audio amplifiers, one might input 60 and 6000 Hz signals; unintentional curvature in the input–output characteristic of the amplifier may cause such multiplication. The presence of new signals at 5940 and 6060 Hz is called "modulation distortion" — sometimes "intermodulation distortion." In the receiver, one looks for a circuit with a pronounced multiplication effect; sometimes the circuit is called a "mixer," sometimes a "frequency converter," perhaps even, if one prefers Greek-rooted words, a "heterodyne circuit." Also, the receiver using these new techniques is called a "heterodyne receiver."

Naturally, one of the two frequencies being fed into the mixer is the carrier frequency one wishes to detect. The other one must be locally generated in the receiver by an oscillator circuit called the "local oscillator," abbreviated LO. Let us now look at desirable choices for local oscillator frequency. Either the *sum* of the IF and the LO must be the carrier frequency to be received or the *difference* of LO and IF must be the carrier frequency. Let us examine some numbers to see which choice of local oscillator is more feasible.

Example 3.3

Let us first consider having IF and LO add up to the carrier being received. Consider first, the low end of the AM band, 530 kHz. The standard choice for IF in AM receivers is 455 kHz. Thus, we would have

$$455 \text{ kHz} + f_{LO} = 530 \text{ kHz}; \therefore f_{LO} = 75 \text{ kHz}$$

Since the local oscillator frequency is *below* the carrier frequency being received, this is called "low-side injection." Next, suppose we wish to receive 1620 kHz. The equation for low-side injection would ask for

$$455 \text{ kHz} + f_{LO} = 1620 \text{ kHz}; \therefore f_{LO} = 1165 \text{ kHz}$$

Now, we can see why low-side injection is so impractical for the AM band. The ratio of highest to lowest oscillator frequencies we must produce locally is about 15 to 1. We will see a bit later that oscillator frequencies usually depend upon the frequency of a resonant circuit, so a variable oscillator depends upon a resonant circuit with variable capacitor or inductor. But because resonance frequency depends upon the *square* root of L and C, low-side injection requires a variation of L or C of about 200 to 1. About the highest variation one can hope to buy is of order 10:1. We had best inspect *high*-side injection.

SOLUTION Now, we try

$$f_{Lo} - 455 \text{ kHz} = 530 \text{ kHz}; \quad \therefore f_{Lo} = 985 \text{ kHz}$$

At the high end of the band, the equation is

$$f_{LO} - 455 \text{ kHz} = 1620 \text{ kHz}; \quad \therefore f_{LO} = 2065 \text{ kHz}$$

Here, the ratio of local oscillator frequencies at top and bottom of the band is barely over 2, requiring capacitance or inductance to vary 4:1. This is achievable with no great effort, so high-side injection is the only feasible option in AM receivers.

EXERCISES 3.3

1. The broadcast band for FM radio in the U.S. assigns carrier frequencies 200 kHz apart, starting at 88.1 MHz and running up to 107.9 MHz. FM receivers use IF amplifiers tuned to 10.7 MHz. Find the range of local oscillator frequencies for low- and high-side injection. Do you see circuit feasibility problems in choosing low side rather than high side?

ANSWERS 77.4 to 97.2 MHz, low side; 98.8 to 118.6 MHz, high side.
No real circuit feasibility problems, as we only require about a 1.25:1 range in local oscillator frequency regardless of whether we choose low- or high-side injection.

2. For channel 2, the video carrier is located at 55.25 MHz. For channel 6, the video carrier is at 83.25 MHz. In TV receivers, the IF for video is at 45.75 MHz. Find the range of local oscillator frequencies required for low- and high-side injection and comment upon your findings.

ANSWERS 9.5 to 37.5 MHz, low side (about a 4:1 ratio); 101 to 129 MHz, high side. Low side looks pretty iffy; high side, pretty easy.

3.4 Image Frequency and RF Amplifier Requirements

Okay, now everything is cool and under control, right? Well, not quite. Suppose we are using the sensible high-side injection to receive 530 kHz by using a local oscillator which puts out a frequency of 985 kHz. Suddenly, a signal having its carrier at 1440 kHz appears at the RF input to the mixer. The mixer dutifully puts out sum and difference frequencies using the new carrier, yielding new carriers at 455 kHz and at 1895 MHz. The second frequency will be largely attenuated by the IF amplifier, but the lower one is exactly right for the IF amplifier, and hence is able to interfere with the desired signal at 530 kHz. The interfering carrier at 1440 kHz is called an "image" of 530 kHz. If the reader looks into a mirror, he/she sees the image of an individual who looks a lot like oneself, apparently standing exactly as far behind the mirror as the individual is in front of it (except if the reader parts hair on the left, the image obviously parts on the right). Similarly, the image frequency is an equal amount above the LO compared to how much the image is below it.

Clearly, images can be quite a problem, unless they are attenuated before they ever reach the mixer. This, then, is the task of the "RF amplifier" if one is included in the receiver. The image problem arose because of use of the mixer to heterodyne the input frequency to a new frequency, including the RF circuit meant to knock down the amplitude of the image making the receiver into a superheterodyne variety. It takes in the image signal and attenuates it rather than passing it on to the mixer, where it becomes a problem. Note that it would take an unrealistically large Q in the RF amplifier to do any of the needed selectivity against a next channel to the one being tuned (a so-called adjacent channel) or one of the channels two channels up or down (a so-called alternate channel). One hopes that the circuit's optimum is *near* the frequency preferred, hoping only that the image is significantly attenuated. "Tracking" can be a slight problem in the receiver. Since the preferred frequency for the RF amplifier to be favored varies over a factor of 3:1, its tuning capacitor would be "ganged" on the same shaft as the local oscillator capacitor, so they both may not reach their optimum at the same time.

The poorest rejection of the RF amplifier against the image frequency occurs when their ratio is lowest, which occurs at the high end of the AM band. Parallel resonance is often worked with as though the circuit itself is made up of purely parallel elements, as shown in Figure 3.4.

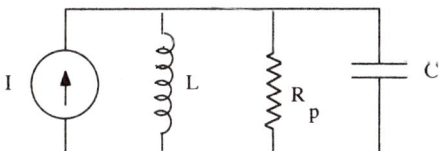

FIGURE 3.4
Purely parallel resonator circuit.

For this circuit, we may write the output voltage as

$$V_o = \frac{1}{\frac{1}{R_p} + j\omega C - \frac{j}{\omega L}} = \frac{IR_p}{1 + jR_p(\omega C - 1/\omega L)} = \frac{IR_p}{1 + jQ(f/f_o - f_o/f)},$$

where, as it was before, $f_o = 1\,1/2\pi\sqrt{LC}$.

Example 3.4

Find the image rejection when the RF signal is 1000 kHz. Assume that the Q of the RF circuit stays constant at 20, is independent of frequency, and that tracking is no problem (the circuit is tuned exactly to the RF frequency, whatever it is).

SOLUTION   When input frequency is 1000 Hz, the local oscillator frequency is 1465 kHz, and the image is at 1930 kHz. Basically, we need to evaluate

$$Q\left(\frac{f}{f_0} - \frac{f_0}{f}\right) \text{ when } f = 1930 \text{ kHz and } f_0 = 1000 \text{ kHz.}$$

We get $20 \times (1.930 - 1/1.930) = 28.24$ which is very much greater than 1, the other term in the denominator. Since the denominator is 1 at resonance, we just take $20 \log_{10}(28.24)$ and get 29.0 dB.

EXERCISES 3.4

1. Find the image rejection when the RF signal is 600 and 1600 kHz for the RF amplifier above.

ANSWERS   32.5 dB, 25.4 dB.

2. It is when one looks at the most likely image signals which might be experienced that one sees the logic in choosing high-side injection in FM receivers. The FM band lies just above channel 6, so the most likely images to be experienced would be the sound carriers of the low TV channels, which might well be stronger than the FM signals one wished to receive. Assume that one was so rash as to choose low-side injection. Find TV sound carriers that would then serve as images in the FM band. (Hint: channel 2 has its sound carrier at 59.75 MHz, that for channel 3 is at 65.75 MHz, channel 4 has sound at 75.75 MHz, channel 5 has it at 81.75 MHz, and channel 6 is at 87.75 MHz.) Don't forget that the intermediate frequency used in FM receivers is 10.7 MHz, so low-side injection would put the local oscillator 10.7 MHz below the FM station being tuned, but 10.7 MHz above the TV sound carrier that might be present in the neighborhood of the receiver.

ANSWERS   To receive channel 2 sound would require LO at 70.45 MHz, which for low-side injection would lead it to interfere with an FM channel at 81.45 MHz, of which there are none, so no problem. Channel 3 sound would be an image of FM signals at 87.45 MHz, which barely misses the low side of the FM band. **However,** channel 4 would have its image at 97.45 MHz, almost dead center in the FM band, and channel 5 would be an image at 103.45 MHz. Channel 6 sound would be an image at 109.45 MHz, which misses the high end of the FM band. Note also that none of these TV stations *is in fact* an FM image problem because high-side injection is always used.

3. There exists a rather substantial gap between channels 6 and 7. Mainly, one gathers, channel 7 was placed high enough that with high-side injection used in TV receivers, its video would not would not fall the same place as the channel 6 video carrier. Show that this is not a problem.

ANSWERS   Channel 6 video at 83.25 MHz has its LO at 129.0 MHz, making an image of 174.75 MHz. Channel 7 video is at 175.25, so we are a bit marginal here.

4. Suppose that one has an AM carrier at 30 MHz, and that one proposes to use an IF amplifier of 455 kHz.

a. Assuming an RF amplifier with Q of 20 is used, how much will it discriminate against the image here?

b. The big problem we have seen regarding image rejection may be attacked through "double conversion," meaning that one uses two different IF amplifiers to help image rejection. Assume the first IF is at 10 MHz, and the second at 455 kHz. Find the new image frequencies (note that the existence of signals at these frequencies is of low probability).

c. Another devious trick which might work, now that one can get prefabricated IF amplifiers that work at high frequencies, is to convert the input upwards. Suppose that a receiver meant to function with carrier frequencies of 1 MHz to 30 MHz, uses an IF amplifier at 40 MHz. Find image frequencies for 1, 15, and 30 MHz and the characteristics of an amplifier which passes the input signals and rejects the images.

ANSWERS Images are at 81, 95, and 110 MHz. All that we need to get rid of images and yet to pass input carriers is a *low-pass* filter with cutoff above 30 MHz and below 81 MHz.

3.5 Piezoelectric Crystal Resonance

Piezoelectric crystals have mechanical vibrations which have very high efficiency, which is to say there is almost no energy lost in a period of oscillation. The result is that there is a very high-Q series resonance. Consider Figure 3.5, showing electrical circuit parameters for one crystal.

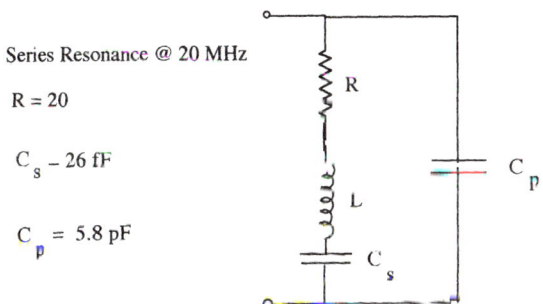

Series Resonance @ 20 MHz

R = 20

C_s – 26 fF

C_p = 5.8 pF

FIGURE 3.5
Equivalent circuit for piezoelectric crystal resonator.

Example 3.5.1

Let's calculate the equivalent value for series inductance and the resulting Q for series resonance.

SOLUTION $L = 1/\omega_0^2 C_s = 1/(4\pi \times 10^7)^2 \times 26 \times 10^{-15} = 2.436$ mH.

Hence, for series resonance, $Q = 2.436 \times 10^{-3} \times 4\pi \times 10^7/20 = 15,306$. This is a Q much larger than will be found anywhere else in low frequency circuits and only can be approached in very carefully made microwave or optical resonators. For this reason, piezoelectric resonators are very useful in building signal sources in which the frequency of operation must be very accurately determined and stable.

Parallel resonance is also obtainable with piezoelectric resonators, but the selectivity, as determined by the Q, will be greatly compromised. The reason is that the frequency of resonance will be determined by the parallel capacitance C_p and the *equivalent* inductance presented just above the series resonant frequency. Since the equivalent inductance will be the series inductance *greatly reduced* by the capacitance in series with it, and series resonance is proportional to inductance whereas R is unchanged and parallel Q is much lowered. Let's do another small calculation.

Example 3.5.2

It is not difficult to prove that at frequencies very near resonance, the reactive impedance of a series-resonant circuit is $2RQ\delta$, where δ is the fractional frequency shift from resonance.

SOLUTION   We first calculate this frequency shift by requiring that the impedance of the equivalent inductor be equal to the reactance of the *parallel* capacitor, so

$$2RQ\delta = 2 \times 20 \times 15,306 \ \delta \cong 1/4\pi \times 10^7 \times 5.8 \times 10^{-12} = 1372; \ \delta = 2.24 \times 10^{-3}.$$

Thus, the frequency of parallel resonance is less than 0.2% higher than series resonance. The *equivalent* inductance is given approximately by

$$L_{eq} \cong 1/(4\pi \times 10^7)^2 \times 5.8 \times 10^{-12} = 10.92 \ \mu H,$$

which is certainly much diminished below the 2.4 mH that teams with the series capacitance to produce series resonance. Using the equivalent inductance but the same resistance as before, we can expect the Q to be approximately

$$Q \cong 4\pi \times 10^7 \times 10.92 \times 10^{-6}/20 = 68.6$$

Thus, one can make parallel resonators using the piezoelectric crystal but it scarcely seems to be recommended, since the Q will be little superior to that with normal inductors and capacitors.

EXERCISES 3.5

1. *Suppose one parallels the resonator above with an extra 50 pF. Estimate the shift in parallel resonant frequency and the parallel Q.*

ANSWERS Frequency raised 0.0233%, Q = 71.

2. *As we saw above, one cannot "pull" the resonant frequency very far from its series value by putting rather large amounts of capacitance in parallel with the resonator. Next, the creative "what–if" engineer might say, "Well, let's put in some series capacitance. That ought to have some effect!"*

ANSWER   If C_x is added in series with the crystal, resonant frequency is

shifted by a factor $\sqrt{\dfrac{C_s + C_0 + C_x}{C_0 + C_x}}$.

3.6   Principles of Sine-Wave Oscillators: The Phase-Shift Oscillator

There are often times when all electrical engineers will need a source of a signal at a specific frequency. This could be supplied by what is called "an oscillator;" an oscillator may be considered an amplifier circuit which supplies an output at a certain frequency without being supplied any input signal. The trick is that one forces the amplifier to supply its own input, through application of appropriate positive feedback. One must connect a circuit called the "feedback network" to the output of the amplifier to supply exactly the input required. Suppose the amplifier boosts the input signal exactly 50 times, but inverts it. Then, the appropriate feedback network reduces the output voltage by a factor of $1/50$ and has a phase shift of 180°. At audio frequencies, this could be three or four cascaded networks, as shown in Figure 3.6.

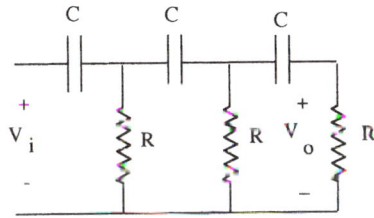

FIGURE 3.6
Phase shift oscillator feedback network.

Suppose we write the impedance of each capacitor as jX. We can then write three mesh equations:

$$I_1(R + jX) - I_2R = V_i,$$

$$-I_1R + I_2(2R + jX) - I_3R = 0, \text{ and}$$

$$-I_2R + I_3(2R + jX) = 0.$$

Because the output voltage depends upon I_3, we simply solve for it in terms of V_i. Then we can write the transfer function relating V_o to V_i as follows:

$$\frac{V_o}{V_i}(j\omega) = 1 - 5\left(\frac{X}{R}\right)^2 + j\left(6\frac{X}{R} - \left(\frac{X}{R}\right)^3\right).$$

The requirement of 180° phase shift will result in the imaginary part of the transfer function required to be zero. Hence, we require $X^2 = 6\,R^2$. This results in the transfer function being equal to –29. Hence, we will have oscillations if we have an inverting amplifier with a gain of at least 29, at such a frequency that

$$\frac{1}{2\pi fC} = \sqrt{6}R.$$

EXERCISES 3.6
1. *Find the value of C required for oscillations at 1 kHz, if R = 4.7 kΩ.*

ANSWER   0.0138 μF.
2. *Repeat step 1 for a frequency of 4 GHz if R = 50Ω is used.*

ANSWER   0.325 pF.
   Besides the difficulty getting the required passive chip components, finding an amplifier with a gain of 29 may be very difficult.

3.7   Colpitts and Hartley Configurations

Colpitts and Hartley configurations both use a parallel resonant circuit to provide the feedback from input to output. The difference is that the two configurations are so arranged that the voltage division is across different circuit elements. In the Hartley, the inductor is tapped so that the voltage fed back is taken across a small portion of the inductor. In the Colpitts, the capacitance of the circuit is contained in a series connection of two capacitors. Consider the key elements of a Hartley circuit, shown in Figure 3.7.
   In this circuit, choice of some of the elements is more critical than others; choice of the resistors is made mainly with dc-bias in mind, although one will

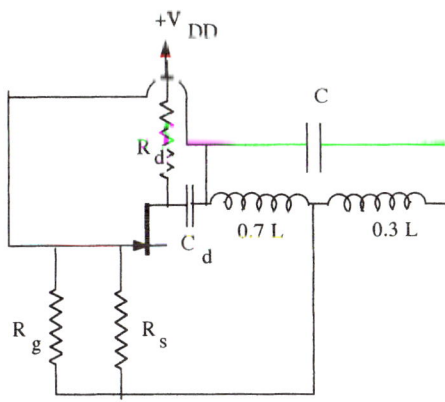

FIGURE 3.7
Hartley oscillator circuit using FET.

try to keep them high enough that they do not appreciably load the resonant circuit and make its Q much lower than would simply depend upon losses in the
resistor. Also, C_d is simply there to keep the left portion of inductor from
grounding the dc value of drain voltage, so is just chosen to have negligible
impedance at the frequency of operation. Although it looks as though two
inductors were chosen, in fact those two elements simply represent one
"tapped" inductor, which is to say, a third connection is made part way from
one end to the other end. The exact point of the tap is not critical; the strategy is
simply to feed back sufficient output voltage to be sure that there is more than
enough to cause oscillations to build up. The choice shown implies that the gain
of the FET is surely at least 2.33, and with extra gain, the waveform distorts, and
the loop gain requirement applies to the *fundamental* of the distorted waveform.
The frequency of operation is then expected to be determined simply from the
resonant frequency determined from L and C. One key consideration in the
Hartley oscillator is that the frequency can be easily "tweaked" if there is a magnetic "slug" in the core of the form on which the inductor is wound; then a small
amount of screwing the slug in or out of the core will change the inductance a
small amount. It might be noted that since the FET inverts the input signal as it
amplifies, the required phase reversal in the feedback network is obtained by
connecting gate and drain to opposite ends of the inductor.

Figure 3.8 shows the Colpitts-connected version of an oscillator using an
opamp used in the inverting configuration so as to have a small amount of
inverting gain. It might be desirable to connect a potentiometer as a variable
resistor for either the input resistor or the feedback resistor in the feedback
path to make the gain slightly variable, so that one can make sure the gain is
sufficient for oscillations to build up. Because we have shown the capacitors
equal, theory says that oscillations should build up if feedback is such that
gain is anything over unity, but Murphy's law is perhaps more operative in
building oscillators than almost any other effort an engineer might under-
take, so one must adjust the adjustable resistor until oscillations begin.

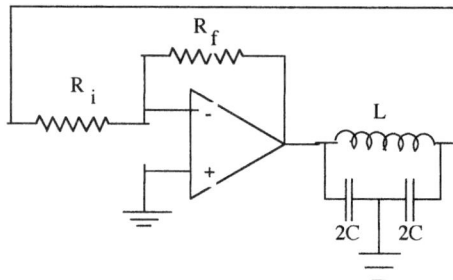

FIGURE 3.8
Opamp-based Colpitts-connected oscillator.

3.8   Crystal-Controlled Oscillators

Use of the piezoelectric crystal resonance as the primary control of the frequency of operation of an oscillator can lead to a very well-defined and stable frequency; because the Q of series resonance is very high, the circuit connections ought to capitalize on this resonance to maximize the frequency stability of the oscillations. One way in which this might be achieved is shown in Figure 3.9.

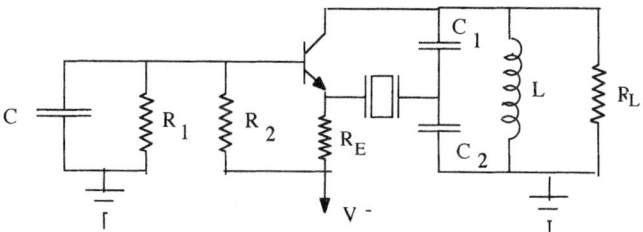

FIGURE 3.9
Crystal-controlled oscillator using feedback to noninverting input.

Working from left to right, we first see a capacitor, which needs simply to be large enough that, at the operation frequency, it is essentially a short-circuit. Hence, we have a common-base configuration, and the input of the amplifier is the emitter. The actual resistors in the circuit are there mainly for the dc-biasing. The two capacitors and the inductor form a parallel resonant circuit, the main purpose of which is to filter out harmonics caused by overdriving the amplifier. It is of course necessary that this resonance be the same frequency as the crystal, but of course its Q is much lower, and the crystal is going to rule the frequency of operation.

3.9   Voltage-Controlled Oscillators

One does not have to go very far in the communications business before one runs across one block in a block diagram labeled "VCO," meaning the title of this section. By "voltage-controlled," what is meant is that the frequency of oscillation depends upon the voltage fed into the VCO input. One way of doing this is fairly elementary to understand. One needs to remember that a reverse-biased P.N. diode acts like a capacitor, the capacitance of which depends upon the amount of the reverse-bias. This diode could be one capacitor of a Colpitts configuration, although it might be desirable to parallel it with a fixed capacitor because using the diode alone might provide too much variation of frequency. If one is considering this method to build a so-called "direct" frequency for the FM broadcast band, one needs remember that the modulation intensity required is a peak variation of frequency of ±75 kHz at a center frequency around 100 MHz, or about 0.075%, requiring capacitance variation of 0.15%.

Example 3.9.1 Use of Varactor Diode in Colpitts VCO

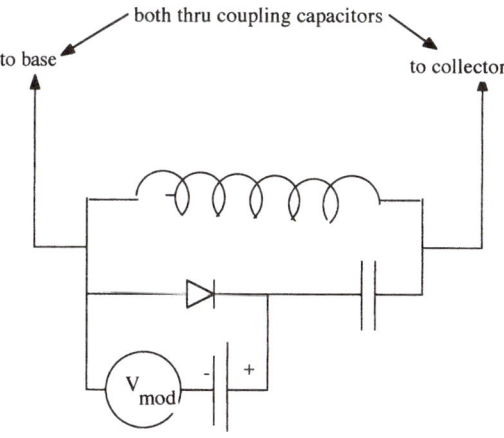

FIGURE 3.10
Varactor diode control of resonance.

Data for the varactor diode from Motorola with the number MMBV 2105 were inspected seeking a linear relation that would describe the variation of capacitance with reverse-bias. When the negative bias is held between 1 and 2 volts, a reasonable linear approximation for the capacitance of the diode is

$$C = (28 - 5v) \text{ pF}.$$

Let us consider capacitance contributed for the varactor diode for each tenth of a volt from 1.0 to 2.0 volts. Combine that capacitance *in series* with 25 pF. Then we will compute resonance frequency for that capacitance in parallel with 100 nH, and from frequency changes from point to point, will get an idea of the frequency modulation sensitivity of the circuit.

v	C	C$_{series}$	f	Δf
1.0	23	11.979	145.416 MHz	
1.1	22.5	11.842	146.254	739
1.2	22.0	11.702	147.126	872
1.3	21.5	11.559	148.033	887
1.4	21.0	11.413	148.977	944
1.5	20.5	11.264	149.959	982
1.6	20.0	11.111	150.968	1009
1.7	19.5	10.955	152.059	1091
1.8	19.0	10.795	153.182	1123
1.9	18.5	10.632	154.352	1170
2.0	18.0	10.465	155.579	1227

EXERCISES 3.9

The linearity of Δf vs. voltage in the example above is not impressive. Try putting the reverse-biased diode in series with 50 pF and repeat the example above:

v	C	C$_{series}$	f	Δf
1.0	23	15.753 pF	126.806 MHz	
1.1	22.5	15.517	127.766	960 Hz
1.2	22	15.278	128.762	996
1.3	21.5	15.035	129.798	1036
1.4	21	14.789	130.873	1075
1.5	20.5	14.539	131.993	1120
1.6	20	14.286	133.157	1164
1.7	19.5	14.029	134.371	1214
1.8	19	13.768	135.639	1268
1.9	18.5	13.504	136.958	1319
2.0	18	13.235	138.343	1385

This is slightly more linear than the example, but not likely to be adequate for any but the least critical applications.

Example 3.9.2 Capacitor-Charging Determination of Frequency

Some of the more sophisticated VCOs determine their operating frequency by charging a capacitor from one voltage toward another, with each half-period ending when the capacitor voltage reaches a certain threshold. To make the calculation a *little* specific, let us choose some numbers. Suppose a capacitor voltage is being charged from –5 volts toward +5 volts. Find the frequency of cooperation in terms of the charging time constant for the capacitor if a semiperiod ends when the capacitor voltage reaches –2 volts.

SOLUTION   From the first circuits course, we know we can write the capacitor voltage as $v_c = 5 - 10\ e^{(-t/\tau)}$ for all time; but in terms of period T, here we
may write

$$-2 = 5 - 10\ e^{(-t/t)};$$

$$10\ e^{-(T/2t)} = 7;$$

$$\ln(10/7) = 0.35667 = T/2\tau$$

$$T = 0.713\ \tau;$$

$$f_{operating} = 1.44018/\tau.$$

EXERCISES 3.9.2
*1. Repeat the example above for switching voltages of –3, –1, 0, 1, 2, 3 and check vari-
ation of operating frequency with voltage.*

ANSWERS

Switching Voltage	Operating Frequency
–3	$2.24/\tau$
–1	$0.979/t$
0	$0.721/\tau$
1	$0.546/\tau$
2	$0.416/\tau$
3	$0.310/\tau$

*These results are not looking very linear in terms of frequency modulation, but we do
find a wide frequency variation of more than 7:1.*

4

Modulation and Demodulation Circuitry

4.1   Some Fundamentals: Why Modulate?

Because this chapter uses a building block approach more than do previous chapters, it may seem to be a long succession of setting up straw men and demolishing them. To some extent, this imitates the development of radio and TV, which has been going on for most of the century now ending. A large number of concepts were developed as the technology advanced; each advance made new demands upon the hardware. At first, many of these advances were made by enthusiastic amateurs who had no fear of failure and viewed radio communication the way Hillary viewed Everest — something to be surmounted "because it was there." Since about World War II, there have been increasing numbers of engineers who understood these principles and could propose problem solutions which might work the first or second time they were tried. The author fondly hopes this book will help to grow a new cadre of problem-solvers for the twenty-first century.

What probably first motivated the inventors of radio was the need for ships at sea to make distress calls. It may be interesting to note that the signal to be transmitted was a digital kind of thing called Morse Code. Later, the medium became able to transmit equally crucial analog signals, such as a soldier warning, "Watch out!! The woods to your left are full of the abominable enemy!" Eventually, during a period without widespread military conflict, radio became an entertainment medium, with music, comedy, and news, all made possible by businessmen who were convinced you could be per-suaded, by a live voice, to buy soap, and later, detergents, cars, cereals not needing cooking, and so on. The essential low and high frequency content of the signal to be transmitted has been very productive for problems to be solved by radio engineers.

The man on radio, urging you to buy a "pre-owned" Cadillac, puts out most of his sound energy below 1000 Hz. A microphone observes pressure fluctuations corresponding to the sound and generates a corresponding volt-age. Knowing that all radio broadcasting is done by feeding a voltage to an antenna, the beginning engineer might be tempted to try sending out the microphone signal directly. A big problem with directly broadcasting such a signal is that an antenna miles long would be required to transmit it efficiently. However, if the frequency of the signal is shifted a good deal higher, effective antennas become much shorter and more feasible to fabricate. This upward translation of the original message spectrum is perhaps the most crucial part of what we have come to call "modulation." However, the necessities of retrieving the original message from the modulated signal may dictate other inclusions in the broadcast signal, such as a small or large voltage at the center, or "carrier" frequency of the modulated signal. The need for a carrier signal is dictated by what scheme is used to transmit the modulated signal, which determines important facts of how the signal can be demodulated.

More perspective on the general problem of modulation is often available by looking at the general form of a modulated signal,

$$f(t) = A(t)\cos\theta(t).$$

If the process of modulation causes the multiplier $A(t)$ out front to vary, it is considered to be some type of "amplitude" modulation. If one is causing the angle to vary, it is said to be "angle" modulation, but there are two basic types of angle modulation. We may write

$$\theta(t) = \omega_c t + \phi(t).$$

If then our modulation process works directly upon $\omega_c = 2\pi f_c$, we say we have performed "frequency" modulation. If, instead, we directly vary the phase factor $\phi(t)$, we say we have performed "phase" modulation. The two kinds of angle modulation are closely related, so that we may do one kind of operation to get the other result, by proper preprocessing of the modulation signal. Specifically, if we put the modulating signal through an integrating circuit before we feed it to a phase modulator, we come out with frequency modulation. This is, in fact, often done. The dual of this operation is possible but is seldom done in practice. Thus, if the modulating signal is fed through a differentiating circuit before it is fed to a frequency modulator, the result will be phase modulation. However, this process offers no advantages to motivate such efforts.

4.2   How to Shift Frequency

Our technique, especially in this chapter, will be to make our proofs as simple as possible; specifically, if trigonometry proves our point, it will be used instead of the convolution theorem of circuit theory. Yet, use of some of the aspects of convolution theory can be enormously enlightening to those who understand. Sometimes, as it will in this first proof, it may also indicate the kind of circuit which will accomplish the task. We will also take liberties with the form of our modulating signal. Sometimes we can be very general, in which case it may be identified as a function m(t). At other times, it may greatly simplify things if we write it very explicitly as a sinusoidal function of time

$$m(t) = \cos \omega_m t.$$

Sometimes, in the theory, this latter option is called "tone modulation," because, if one listened to the modulating signal through a loudspeaker, it could certainly be heard to have a very well-defined "tone" or pitch. We might justify ourselves by saying that theory certainly allows this, because any particular signal we must deal with could, according the theories of Fourier, be represented as a collection, perhaps infinite, of cosine waves of various phases. We might then assess the maximum capabilities of a communication system by choosing the highest value that the modulating signal might have. In AM radio, the highest modulating frequency is typically about $f_m = 5000$ Hz. For FM radio, the highest modulation frequency might be $f_m = 19$ kHz, the frequency of the so-called FM stereo "pilot tone."

In principle, the shifting of a frequency is very simple. This is fairly obvious to those understanding convolution. One theorem of system theory says that multiplication of time functions leads to convolution of the spectra. Let us just multiply the modulating signal by a so-called "carrier" signal. One is allowed to have the mental picture of the carrier signal "carrying" the modulating signal, in the same way that homing pigeons have been used in past wars to carry a light packet containing a message from behind enemy lines to the pigeon's home in friendly territory. So, electronically, for "tone modulation," we need only to accomplish the product

$$\phi(t) = A \cos \omega_m t \cos \omega_c t.$$

Now, we may enjoy the consequences of our assumption of tone modulation by employing trigonometric identities for the sum or difference of two angles:

$$\cos(A + B) = \cos A \cos B - \sin A \sin B \text{ and } \cos(A - B) = \cos A \cos B + \sin A \sin B$$

If we add these two expressions and divide by two, we get the identity we need:

$$\cos A \cos B = 0.5 \left[\cos(A + B) + \cos(A - B) \right].$$

Stated in words, we might say we got "sum and difference frequencies," but neither of the original frequencies. Let's be just a little more specific and say we started with f_m = 5000 Hz and f_c = 1 MHz, as would happen if a radio station whose assigned carrier frequency was 1 MHz were simply transmitting a single tone at 5000 Hz. In "real life," this would not be done very often, but the example serves well to illustrate some definitions and principles. The consequence of the mathematical multiplication is that the new signal has two new frequencies at 995 kHz and 1005 kHz. Let's now just add one modulating tone at 3333 Hz. We would have added two frequencies at 9666.667 kHz and 1003.333 kHz. However, if this multiplication was done purely, *there is no carrier frequency term present*. For this reason, we say we have done a type of "suppressed carrier" modulation. Also, furthermore, we have two *new* frequencies for *each* modulating frequency. We define all of those frequencies above the carrier as the "upper sideband" and all the frequencies below the carrier as the "lower sideband." The whole process we have done here is named "double sideband suppressed carrier" modulation, often known by its initials DSB–SC. Communication theory would tell us that the signal spectrum, before and after modulation with a single tone at a frequency f_m, would appear as in Figure 4.1. Please note that the theory predicts equal positive and negative frequency components. There is no deep philosophical significance to negative frequencies. They simply make the theory symmetrical and a bit more intuitive.

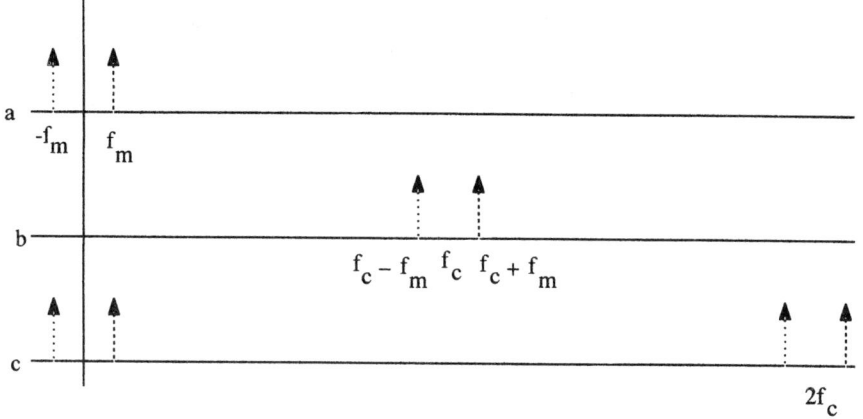

FIGURE 4.1
Unmodulated, modulated, and synchronously demodulated signal spectra. a. Spectrum of tone-modulating signal. b. Spectrum (positive part only) of double sideband suppressed carrier signal. c. Spectrum of synchronously detected DSB-SC signal (except for part near $-2f_c$).

4.3 Analog Multipliers, or "Mixers"

First, there is an unfortunate quirk of terminology; the circuit that multiplies signals together is in communication theory usually called a "mixer." What

is unfortunate is that the engineer or technician who produces sound record-
ings is very apt to feed the outputs of many microphones into potentiome-
ters, the outputs of which are sent in varying amounts to the output of a piece
of gear, and *that* component is called a "mixer." Thus, the communication
engineer's mixer multiplies and the other adds. Luckily, it will usually be
obvious which device one is speaking of.

There are available a number of chips (integrated circuits) designed to
serve as analog multipliers. The principle is surprisingly simple, although the
chip designers have added circuitry which no doubt optimizes the operation
and perhaps makes external factors less influential. The reader might remember that the transconductance g_m for a bipolar transistor is proportional to the
collector current; its output is proportional to the g_m *and* the input voltage, so
in principle one can replace an emitter resistor with the first transistor, which
then controls the collector current of the second transistor. If one seeks to fabricate such a circuit out of discrete transistors, one would do well to expect a
need to tweak operating conditions considerably before some approximation
of analog multiplication occurs. Recommendation: buy the chip. Best satis-
faction will probably occur with a "four-quadrant multiplier." The alternative is a "two-quadrant multiplier," which might embarrass one by being
easily driven into cut-off.

Another effective analog multiplier is alleged to be the dual-gate FET. The
width of the channel in which current flows depends upon the voltage on
each of two gates which are insulated from each other. Hence, if different
voltages are connected to the two gates, the current that flows is the product
of the two voltages. Both devices we have discussed so far have the advantage of having some amplification, so the desired resulting signal has a
healthy amplitude. A possible disadvantage may be that spurious signals one
does *not* need may also have strong amplitudes.

Actually, the process of multiplication may be the byproduct of any distorting amplifier. One can show this by expressing the output of a distorting
amplifier as a Taylor series representing output in terms of input. In principle, such an output would be written

$$V_o = a_0 + a_1(v_1 + v_2) + a_3(v_1 + v_2)^2 + \text{smaller terms.}$$

One can expand $(v_1 + v_2)^2$ as $v_1^2 + 2v_1v_2 + v_2^2$, so this term yields second harmonic terms of each input plus the product of inputs one was seeking. However, the term $a_1(v_1 + v_2)$ also yielded each input, so the carrier here would not
be suppressed. If it is fondly desired to suppress the carrier, one must resort
to some sort of "balanced modulator." An "active" (meaning there is amplification provided) form of a balanced modulator may be seen in Figure 4.2;
failure to bias the bases of the transistors should assure that the voltage
squared term is large.

One will also find purely passive mixers with diodes connected in the
shape of a baseball diamond with one signal fed between first and third base,

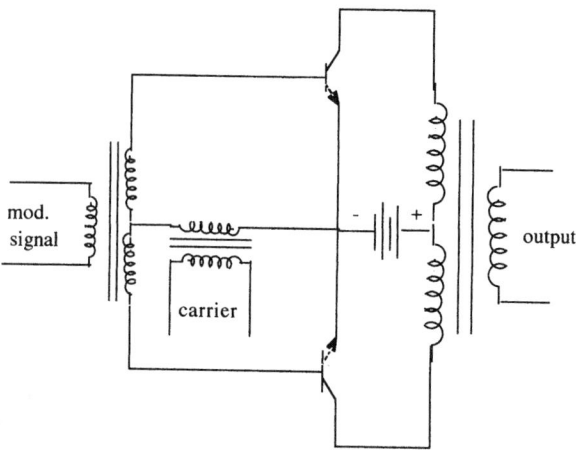

FIGURE 4.2
Balanced modulator.

the other from second to home plate. Such an arrangement has the great advantage of not requiring a power supply; the disadvantage is that the amplitude of the sum or difference frequency may be small.

4.4   Synchronous Detection of Suppressed Carrier Signals

At this point, the reader without experience in radio may be appreciating the mathematical tricks but wondering, if one can accomplish this multiplication, can it be broadcast and the original signal retrieved by a receiver? A straightforward answer might be that multiplying the received signal by another carrier frequency signal such as $\cos \omega_c t$ will shift the signal back exactly to where it started and also up to a center frequency of twice the original carrier. This is depicted in part c of Figure 4.1. The name of this process is "synchronous detection." (In the days when it was apparently felt that communications enjoyed a touch of class if one used words having Greek roots, they called it "homodyne detection." If the reader reads a wide variety of journals, he/she may still encounter the word.) The good/bad news about synchronous detection is that the signal being used in the detector multiplication must have the *exact frequency and phase* of the original carrier, and such a signal is not easy to supply. One method is to send a "pilot" carrier, which is a small amount of the correct signal. The pilot tone is amplified until it is strong enough to accomplish the detection.

Suppose the pilot signal reaches high enough amplitude but is phase-shifted an amount θ with respect to the original carrier. We would then in our synchronous detector be performing the multiplication:

$$m(t)\cos\omega_c t\,\cos(\omega_c t + \theta).$$

To understand what we get, let us expand the second cosine using the identity for the sum of two angles,

$$\cos(\omega_c t + \theta) = \cos\omega_c t\cos\theta - \sin\omega_c t\sin\theta.$$

Hence, the output of the synchronous detector may be written as

$$m(t)\cos^2\omega_c t\cos\theta - m(t)\cos\omega_c t\sin\omega_c t\sin\theta =$$

$$(0.5)\big[m(t)\cos\theta(1-\cos 2\omega_c t) - m(t)\sin\theta\sin 2\omega_c t\big].$$

The latter two terms can be eliminated using a low-pass filter, and one is left with the original modulating signal, m(t), attenuated proportionally to the factor cos θ, so major attenuation does not appear until the phase shift approaches 90°, when the signal would vanish completely. Even this is not totally bad news, as it opens up a new technique called "quadrature amplitude modulation."

The principle of QAM, as it is abbreviated, is that entirely different modulating signals are fed to carrier signals that are 90° out of phase; we could call the carrier signals cos ω_c t and sin ω_c t. The two modulating signals stay perfectly separated if there is no phase shift to the carrier signals fed to the synchronous detectors. The color signals in a color TV system are QAM'ed onto a 3.58 MHz subcarrier to be combined with the black-and-white signals, after they have been demodulated using a carrier generated in synchronism with the "color burst," (several periods of a 3.58 MHz signal) which is cleverly "piggy-backed" onto all the other signals required for driving and synchronizing a color TV receiver.

EXERCISES 4.4
Prove that if the carrier signals fed to the synchronous detectors in a QAM system are both phase-shifted an amount θ, not only is the desired signal attenuated an amount cos θ, but the undesired signal will be present in an amount proportional to sin θ. The ratio of undesired to desired signal, which might be called a "crosstalk ratio," will then be tan θ. Find the phase error for crosstalk ratios of 10, 20, 30, and 40 dB.

ANSWERS 17.55°, 5.71°, 1.81°, 0.57°.

4.5   Single Sideband Suppressed Carrier

The alert engineering student may have heard the words "single sideband" and be led to wonder if we are proposing sending one more sideband than necessary. Of course it is true, and SSB–SC, as it is abbreviated, is the method of choice for "hams," the amateur radio enthusiasts who love to see night fall, when their low wattage signals can bounce between the earth and a layer of ionized atmospheric gasses 100 or so miles up until they have reached half-way around the world. It turns out that a little phase shift is not a really drastic flaw for voice communications, so the "ham" just adjusts the variable frequency oscillator being used to synchronously demodulate incoming signals until the whistles and squeals become coherent, and then he/she listens.

How can one produce single sideband? For many years it was pretty naïve to say, "Well, let's just filter one sideband out!" This would have been very naïve because, of course, one does not have textbook filters with perfectly sharp cut-offs. Recently, however, technology has apparently provided rather good "crystal lattice filters" which are able fairly cleanly to filter the extra sideband. In general, though, the single sideband problem is simplified if the modulating signal does not go to really deep low frequencies; a microphone that does not put out much below 300 Hz might have advantages, as it would leave a transition region of 600 Hz between upper and lower sidebands in which the sideband filter could have its amplitude response "roll off" without letting through much of the sideband to be discarded. Observe Figure 4.3, showing both sidebands for a baseband signal extending only from 300 Hz to 3.0 kHz.

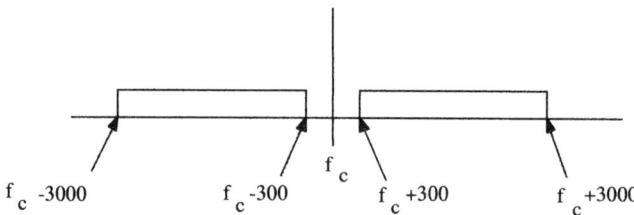

FIGURE 4.3
Double sideband spectrum for modulating signal 300–3000 Hz.

Another method of producing single sideband, called the "phase-shift method," is suggested if one looks at the mathematical form of just one of the sidebands resulting from tone modulation. Let us just look at a lower sideband. The mathematical form would be

$$v(t) = A \cos(\omega_c - \omega_m)t = A \cos \omega_c t \cos \omega_m t + A \sin \omega_c t \sin \omega_m t$$

Mathematically, one needs to perform DSB–SC with the original carrier and modulating signals (the cosine terms) and also with the two signals each phase shifted 90°; the resulting two signals are then added to obtain the lower sideband. Obtaining a 90° phase shift is not difficult with the carrier, of which there is only one, but we must be prepared to handle a band of modulating signals, and it is not an elementary task to build a circuit that will produce 90° phase shifts over a range of frequency. However, a reasonable job will be done by the circuit of Figure 4.4 when the frequency range is limited (e.g., from 300 to 3000 Hz). Note that one does *not* modulate directly with the original modulation signal, but that the network uses each input frequency to generate two signals which are attenuated equal amounts and 90° away from each other. These voltages would be designated in the drawing as V_{xz} and V_{yz}. In calculating such voltages, the reader should note that there are two voltage dividers connected across the modulating voltage, determining V_x and V_y, and that from both of these voltages is subtracted the voltage from the center-tap to the bottom of the potentiometer. Note also that the resistance of the potentiometer is not relevant as long as it does not load down the source of modulating voltage, and that a good result has been found if the setting of the potentiometer is for 0.224 of the input voltage.

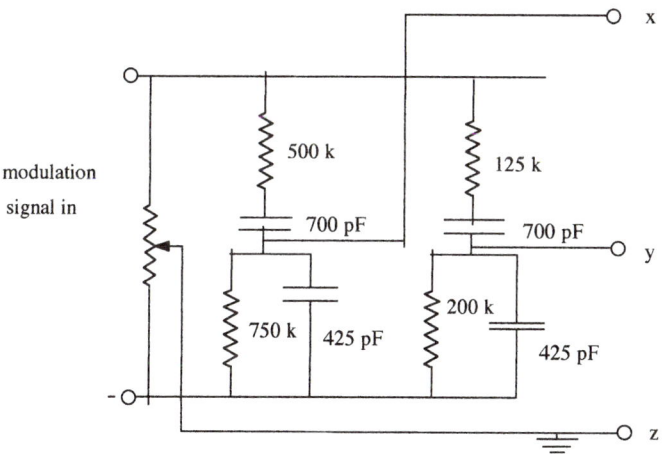

FIGURE 4.4
Audio network for single sideband modulator.

EXERCISES 4.5
Find V_{xz} and V_{yz} at a number of frequencies in the range 300-3000 Hz, say for end-points plus 600, 1000, 1500, 2000, and 2500 Hz.

SOME ANSWERS (Numbers are fraction and phase of input voltage at each frequency.)

$$V_{xz} = 0.216\angle30.58°, \ V_{yz} = 0.224\angle117.65° \text{ at 300 Hz.}$$

$$V_{xz} = 0.216\angle-15.05°, \ V_{yz} = 0.224\angle74.40° \text{ at 600 Hz.}$$

$$V_{xz} = 0.224\angle59.50°, \ V_{yz} = 0.224\angle148.84° \text{ at 300 Hz.}$$

4.6   Amplitude Modulation as Double Sideband with Carrier

The budding engineer must understand that synchronous detectors are more expensive than many people can afford, and that a less expensive detection method is needed. What fills this bill much of the time is called the "envelope detector." Let us examine some waveforms, first for DSB–SC and then for a signal having a large carrier component. Figure 4.5 shows a waveform in which not very different carrier and modulating frequencies were chosen so that a spreadsheet plot would show a few details.

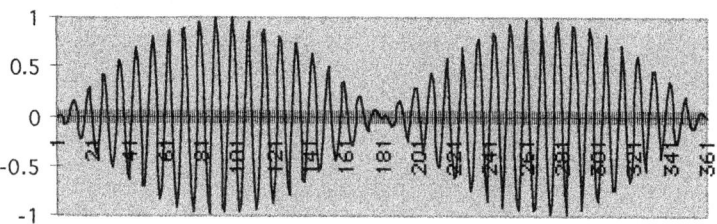

FIGURE 4.5
Double sideband suppressed carrier signal.

An ideal circuit we call an envelope detector would follow the topmost excursion of the waveform sketched here. Now, the original modulating signal was a sine wave, but the topmost excursion would be a *rectified* sinusoid, thus containing large amounts of harmonic distortion. How can one get a waveform that will be detected without distortion by an envelope detector? What was plotted was $1.0 \cos \omega_c t \cos \omega_m t$. We suspect we must add some amount of carrier $B \cos \omega_c t$. The sum will be

$$\phi_{AM}(t) = B \cos \omega_c t + 1.0 \cos \omega_c t \cos \omega_m t = \cos \omega_c t[B + 1.0 \cos \omega_m t].$$

This result is what is commonly called "amplitude modulation." Perhaps the most useful way of writing the time function for an amplitude modulation signal having tone modulation at a frequency f_m follows:

$$\phi_{AM}(t) = A \cos \omega_c t[1 + a \cos \omega_m t].$$

In this expression, we can say that A is the peak amplitude of the carrier signal that would be present if there were no modulation. The total expression inside the [ ] brackets can be called the "envelope" and the factor "a" can be called the "index of modulation." As we have written it, if the index of modulation were >1, the envelope would attempt to go negative; this would make it necessary, for distortion-free detection, to use synchronous detection. "a" is often expressed as a percentage, and when the index of modulation is less than 100%, it is possible to use the simplest of detectors, the envelope detector. We will look at the envelope detector in more detail a bit later.

4.7 Modulation Efficiency

It is good news that sending a carrier along with two sidebands makes inexpensive detection using an envelope detector possible. The accompanying bad news is that the presence of carrier does not contribute *at all* to useful signal output; the presence of a carrier only leads after detection to dc, which may be filtered out at the earliest opportunity. Sometimes, as in video, the dc is needed to set the brightness level, in which case dc may need to added back in at an appropriate level.

To express the effectiveness of a communication system in establishing an output signal-to-noise ratio, it is necessary to define a "modulation efficiency," which, in words, is simply the fraction of output power that is put into sidebands. It is easily figured if the modulation is simply one or two purely sinusoidal tones; for real-life modulation signals, one may have to express it in quantities that are less easy to visualize.

For tone modulation, we can calculate modulation efficiency by simply evaluating the carrier power and the power of all sidebands. For tone modulation, we can write:

$$\phi_{AM}(t) = A\cos\omega_c t\left[1 + a\cos\omega_m t\right] =$$
$$A\cos\omega_c t + (aA)/2\left[\cos(\omega_c + \omega_m)t + \cos(\omega_c - \omega_m t)\right].$$

Now, we have all sinusoids, the carrier, and two sidebands of equal amplitudes, so we can write the average power in terms of peak amplitudes as:

$$P = 0.5\left[A^2 + 2\times(aA/2)^2\right] = 0.5A^2\left[1 + a^2/2\right].$$

Then modulation efficiency is the ratio of sideband power to total power, for modulation by a single tone with modulation index "a," is:

$$\eta = \frac{(aA/2)^2}{0.5A^2(1 + a^2/2)} = \frac{a^2}{2 + a^2}.$$

Of course, most practical modulation signals are not so simple as sinusoids. It may be necessary to state how close one is to overmodulating, which is to say, how close to negative modulating signals come to driving the envelope negative. Besides this, what is valuable is a quantity we shall just call "m," which is the ratio of average power to peak power for the modulation function. For some familiar waveforms, if the modulation is sinusoidal, m = 1/2. If modulation were a symmetrical square wave, m = 1.0; any kind of symmetrical triangle wave has m = 1/3. In terms of m, the modulation efficiency is

$$\eta = \frac{ma^2}{1 + ma^2}$$

EXERCISES 4.7
1. Find the modulation efficiency for 100% modulation using symmetrical square wave, sine wave, and triangle wave.

ANSWERS　50%, 33%, and 25%.
2. If the modulated signal is kept as in Figure 4.5, how much unmodulated carrier must be added to produce modulation indices of 80% and 50%?

ANSWERS　1.25 sin ω_ct, 2.0 sin ω_ct.
3. Suppose that one has an AM signal centered at 560 MHz, with modulation index of 0.5 that one undertakes to amplify using a Class C tuned amplifier with a Q = 40. What would be the modulation index at the output of the amplifier if the modulation frequency is

　　a. 5 kHz?
　　b. 3 kHz?

ANSWERS　40.75%, 45.98%.

4.8　The Envelope Detector

Much of the detection of modulated signals, whether the signals began life as AM or FM broadcast signals or the sound or the video of TV, is done using envelope detectors. Figure 4.7 shows the basic circuit configuration.

The input signal is of course as shown in Figure 4.6. It is assumed that the forward resistance of the diode is 100 ohms or less. Thus, the capacitor is small enough that it gets charged up to the peak values of the high frequency signal, but then when input drops from the peak, the diode is reverse-biased so the capacitor can only discharge through R. This discharge voltage is of course given by

$$V(0)\exp(-t/RC).$$

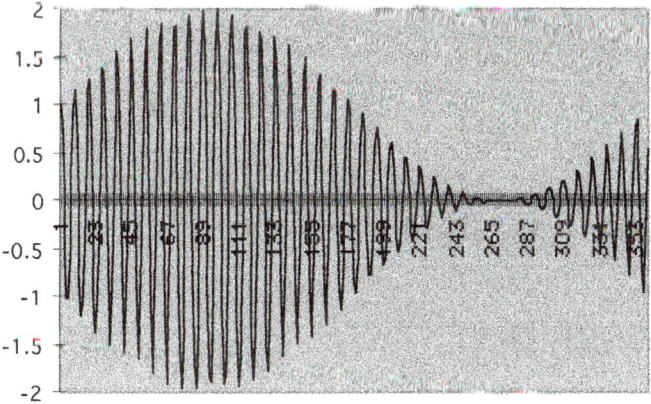

FIGURE 4.6
Amplitude-modulated signal.

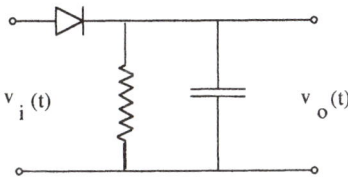

FIGURE 4.7
Simple envelope detector schematic.

Now the problem in AM detection is that we must have the minimum rate of decay of the voltage be at least the maximum decay of the envelope of the modulated wave. We might write the envelope as a function of time:

$$E(t) = A(1 + a \cos \omega_{mt} t),$$

where A is the amplitude of the carrier before modulation and "a" is the index of modulation, which must be less than one for accurate results with the envelope detector. Then, when we differentiate, we get

$$\frac{dE}{dt} = -\omega_m Aa \sin(\omega_m t).$$

We want this *magnitude* to be less than or equal to the maximum magnitude of the rate of decay of a discharging capacitor, which is $E(0)/RC$. For what is written as $E(0)$, we will write the instantaneous value of the envelope, and the expression becomes

$$A(1 + a \cos \omega_{mt} t) \geq RC(\omega_m aA \sin(\omega_m t)).$$

The As cancel, and we have

$$RC \leq \frac{1 + a\cos(\omega_m t)}{\omega_m a \sin(\omega_m t)};$$

our major difficulty occurs when the right-hand side has its minimum value. If we differentiate with respect to $\omega_m t$, we get

$$\frac{\omega_m a \sin\left(\omega_m t \times \left(-\omega_m a \sin(\omega_m t) - \left(1 + a\cos(\omega_m t)\right)(\omega_m)^2 a\cos\omega_m t\right)\right)}{\left(\omega_m a \sin(\omega_m t)\right)^2}.$$

We set the numerator equal to zero to find its maximum. We find we have

$$-(\omega_m a)^2\left[\sin^2(\omega_m t) + \cos^2(\omega_m t)\right] - a(\omega_m)^2 \cos\omega_m t$$

$$= -(\omega_m a)^2 - a(\omega_m)^2 \cos\omega_m t = 0.$$

Hence, the maximum occurs when $\cos \omega_m t = -a$, and of course by identity, at that time, $\sin\omega_m t = \sqrt{1 - a^2}$.

Inserting these results into our inequality for the time constant RC, we have

$$RC \leq \frac{1 - a^2}{\omega_m a\sqrt{1 - a^2}} = \frac{\sqrt{1 - a^2}}{\omega_m a}.$$

Example 4.8

Suppose we say 2000 Hz is the main problem in our modulation scheme, our modulation index is 0.5, and we choose R = 10k to make it large compared to the diode forward resistance, but not *too* large. What should be the capacitor C?

SOLUTION   We use the equality now and get

$$C = \frac{\sqrt{1 - 0.5^2}}{0.5 \times 4000\pi \times 10,000} = 13.8 \text{ nF}.$$

EXERCISES 4.8

1. In demodulating a video picture, maximum modulation of 8% occurs at 15,750 Hz. Our resistor is 20k. Find the maximum size for the capacitor.

ANSWER   6.29 nF.

4.9 Envelope Detection of SSB Using Injected Carrier

Single sideband, it might be said, is a very forgiving medium. Suppose that one were attempting synchronous detection using a carrier that was off by a Hertz or so, compared to the original carrier. Because synchronous detection works by producing sum and difference frequencies, 1 Hz error in carrier frequency would produce 1 Hz error in the detected frequency. Because SSB is mainly used for speech, it would be challenging indeed to find anything wrong with the reception of a voice one has only ever heard over a static-ridden channel. Similar results would also be felt in the following, where we add a carrier to the sideband and find that we have AM, albeit with a small amount of harmonic distortion.

Example 4.9

Starting with just an upper sideband $B \cos(\omega_c + \omega_m)t$, let us add a carrier term $A \cos \omega_c t$, manipulate the total, and prove that we have an envelope to detect. First we expand the sideband term as

$$\Phi_{SSB}(t) = B\left[\cos \omega_c t \cos \omega_m t - \sin \omega_c t \sin \omega_m t\right].$$

Adding the carrier term $A \cos \omega_c t$ and combining like terms, we have

$$\phi(t) = \cos \omega_c t\left(A + B \cos \omega_m t\right) - B \sin \omega_c t \sin \omega_m t.$$

In the first circuits class, we see that if we want to write a function of one frequency in the form $E(t) \cos(\omega_c t + \text{phase angle})$, the amplitude of the multiplier E is the square root of the squares of the coefficients of $\cos \omega_c t$ and $\sin \omega_c t$. Thus,

$$E(t) = \sqrt{\left(A + B \cos \omega_m t\right)^2 + \left(B \sin \omega_m t\right)^2}$$

$$= \sqrt{A^2 + 2AB \cos \omega_m t + B^2\left(\cos^2 \omega_m t + \sin^2 \omega_m t\right)}.$$

Now, of course, the coefficient of B^2 is unity for all values of $\omega_m t$. We find that best performance occurs if $B \ll A$. Then we would have our expression for the envelope (and thus it is detectable using an envelope detector):

$$E(t) = \sqrt{A^2 + B^2 + 2AB \cos \omega_m t} = \sqrt{A^2 + B^2}\sqrt{1 + \frac{2AB}{A^2 + B^2} \cos \omega_m t}.$$

Our condition that $B \ll A$ allows us to say the coefficient of $\cos \omega_m t$ is really small compared to unity. We use the binomial theorem to approximate the second square root: $(1 + x)^n \approx 1 + x/2 + (1/2)^2 (-1/2)x^2$ when $x \ll 1$. Using our approximation, $x \approx (2B/A) \cos \omega_m t$. In our expansion, the x term is the modulation term we were seeking, the x^2 term contributes second harmonic distortion. Using the various approximations, and stopping after we find the second harmonic (other harmonics *will* be present, of course, but in decreasing amplitudes), we have

$$\text{Detected } f(t) = B \cos \omega_m t - (1/2)(B^2/A) \cos^2(\omega_m t).$$

When we use trig identities to get the second harmonic, we get another factor of one half; the ratio of detected second harmonic to fundamental is thus $(1/4)(B/A)$. Thus, for example, if B is just 10% of A, second harmonic is only 2.5% of fundamental.

EXERCISES 4.9
Use the approximate formula derived above to find the conditions for 5% second harmonic and 1%.

ANSWERS $B = 0.2A$, $B = 0.05A$.

4.10   Direct vs. Indirect Means of Generating FM

Let us first remind ourselves of basics regarding FM. We can write the time function in its simplest form as

$$\phi_{FM}(t) = A \cos(\omega_c t + \beta \sin \omega_m t).$$

Now, the alert reader might be saying, "Hold on! That looks a lot like phase modulation. If $\beta = 0$, the phase would increase linearly in time, as an unmodulated signal, but gets advanced or retarded a maximum of β." One needs to remember the definition of instantaneous frequency, which is

$$f_i = \frac{1}{2\pi} \frac{d}{dt}(\omega_c t + \beta \sin \omega_m t) = \frac{1}{2\pi}(2\pi f_c + \beta 2\pi f_m \cos \omega_m t) = f_c + \beta f_m \cos \omega_m t.$$

Thus, we can say that instantaneous frequency departs from the carrier frequency by a maximum amount βf_m, which is the so-called "frequency deviation." This has been specified as a maximum of 75 kHz for commercial FM radio but 25 kHz for the sound of TV signals.

Now, certainly, the concept of directly generating FM has an intellectual appeal to it. The problems of direct FM are mainly practical; if the very means

of putting information onto a high frequency "carrier" is in varying the frequency, it perhaps stands to reason the center value of the operating frequency will not be well nailed down. Direct FM could be accomplished as in
Figure 4.8(a),

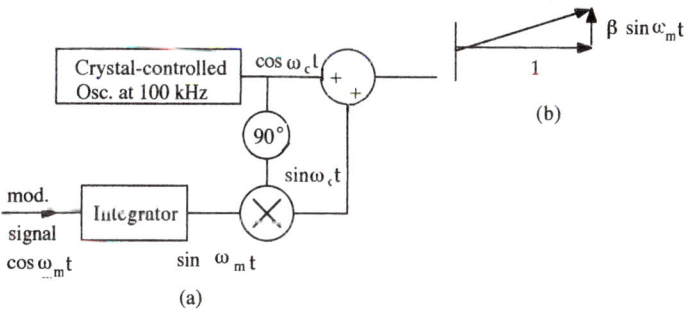

(a)

FIGURE 4.8
(a) Crystal-stabilized phase modulator; (b) phasor diagram.

but Murphy's law would be very dominant and one might expect center frequency to drift continually in one direction all morning and the other way all
afternoon, or the like. This system is sometimes stabilized by having an FM
detector called a discriminator tuned to the desired center frequency, so that
output would be positive if frequency got high and negative for frequency
low. Thus, instantaneous output could be used as an error voltage with a long
time constant to push the intended center frequency toward the center,
whether the average value is above or below.

The best known method of indirect FM gives credit to the man who, more
than any other, saw the possibilities of FM and that its apparent defects could
be exploited for superior performance, Edwin Armstrong. He started with a
crystal-stabilized oscillator around 100 kHz, from which he obtained also a
90° phase-shifted version. A block diagram of just this early part of the Armstrong modulator is shown in Figure 4.8(a).

The modulating signal is passed through an integrator before it goes into an
analog multiplier, to which is also fed the *phase-shifted* version of the crystal-
stabilized signal. Thus, we feed $\cos \omega_c t$ and $\sin \omega_c t \sin \omega_m t$ into a summing
amplifier. The phasor diagram shows the two signals with $\cos \omega_c t$ as the reference. There is a small phase shift given by $\tan^{-1}(\beta \sin \omega_m t)$ where β here gives
the maximum amount of phase shift as a function of time. To see how good a
job we have done, we need to expand $\tan^{-1}(x)$ in a Taylor series. We find that

$$\tan^{-1}(x) = x - (x)^3/3 + (x)^5/5.$$

We see that we have a term proportional to the modulating signal (x) and
others that must represent odd-order harmonic distortion, if one accounts for

the fact that we have resorted to a subterfuge, using a phase modulator to produce frequency modulation. Assuming that our signal finally goes through a frequency detector, we find that the amount of third harmonic as a fraction of the signal output is $\beta^2/4$. Now, in frequency modulation, the maximum amount of modulation which is permitted is in terms of frequency deviation, an amount of 75 kHz. The relation between frequency deviation and maximum phase shift is

$$\Delta f = \beta f_m,$$

where Δf is the frequency deviation, β is maximum phase shift, and f_m is modulation frequency. Since maximum modulation is defined in terms of Δf, the maximum value of β permitted will correspond to *minimum* modulation frequency. Let us do some numbers to illustrate this problem.

Example 4.10

Suppose we have a high fidelity broadcaster wishing to transmit bass down to 50 Hz with maximum third harmonic distortion of 1%. Find the maximum values of β and Δf.

SOLUTION   We have $\beta^2/4 = 0.01$. Solving for β, we get $\beta = 0.2$.
   Then, $\Delta f = 0.2 \times 50$ Hz $= 10$ Hz.
   One can recall that the maximum value of frequency deviation allowed in the FM broadcast band is 75 kHz. Thus, use of the indirect modulator has given us much lower frequency deviation than is allowed, and clearly some kind of desperate measures are required. Such are available, but do complicate the process greatly. Suppose we feed the modulated signal into an amplifier which is not biased for low distortion, that is, its Taylor series looks like

$$a_1 x + a_2 x^2 + a_3 x^3, \text{ etc.}$$

Now the squared term leads to second harmonic, the cubed one gives third harmonic, and so on. The phase-modulated signal looks like $A \cos(\omega_c t + \beta \sin \omega_m t)$ and the term $a_2 x^2$ *not only doubles the carrier frequency, but also the maximum phase shift* β. Thus, starting with the rather low frequency of 100 kHz, we have a fair amount of multiplying room before we arrive in the FM broadcast band 88 to 108 MHz. Unfortunately, we may need different amounts of multiplication for the carrier frequency than we need for the depth of modulation. Let's carry on our example and see the problems that arise. First, if we wish to go from

$$\Delta f = 10 \text{ Hz to } 75,000 \text{ Hz,}$$

that leads to a total multiplication of $75,000/10 = 7500$.

The author likes to say we are limited to frequency doublers and triplers. Let's use as many triplers as possible; we divide the 7500 by 3 until we get close to an even power of 2:

$$7500/3 = 2500; \quad 2500/3 = 833, \quad 833/3 = 278, \quad 278/3 \approx 93, \quad 93/3 = 31,$$

which is very close to $32 = (2)^5$.

So, to get our maximum modulation index, we need five each triplers and doublers. However, 7500×0.1 MHz $= 750$ MHz, and we have missed the broadcast band by about 7 times. One more thing we need is a mixer, after a certain amount of multiplication. Let's use all the doublers and one tripler to get a multiplication of $32 \times 3 = 96$, so the carrier arrives at 9.6 MHz. Suppose our final carrier frequency is 90.9 MHz, and because we have remaining to be used a multiplication of $3^4 = 81$, what comes out of the mixer must be

$$90.9/81 - 1.122 \text{ MHz}.$$

To obtain an output of 1.122 MHz from the mixer, with 9.6 MHz going in, we need a local oscillator of either 10.722 or 8.478 MHz. Note that this local oscillator needs a crystal control also, or the eventual carrier frequency will wander about more than is allowed.

EXERCISES 4.10
1. The station engineer decides that a maximum third harmonic distortion of 4% at 50 Hz is permissible. What would be the key numbers for an Armstrong modulator?

ANSWER   Total multiplication of 3750 needed, or 5 triplers and 4 doublers. Perhaps, the first bit of multiplication might be $3^4 = 81$. That leaves a multiplication of 48, so one needs 1.89375 MHz out of the mixer, thus *crystal-controlled* local oscillator must be at 9.99375 MHz or 6.20625 MHz.

2. Meanwhile, the cynical engineer designing an Armstrong modulator for the station at 107.9 MHz says, "Oh, what the hey. Our listeners are kids without decent woofers anyway. Let's design for 5% third harmonic at 100 Hz!" What are his/her numbers?

ANSWERS $\beta - 0.447$, so $\Delta f - 44.7$ Hz, total multiplication 1677, approximated by $3^3 \times 2^6 = 1728$. Before mixer, try $27 \times 4 = 108$, so 10.8 MHz could be one input into mixer. Output must be 6.47375 MHz, so LO must be 17.54 MHz, or at 4.56 MHz.

4.11   Quick-and-Dirty FM Slope Detection

A method of FM detection that is barely respectable, but surprisingly effective, is called "slope detection." The principle is to feed an FM signal into a tuned circuit, not right at the resonant frequency but rather somewhat off the peak. Therefore, the frequency variations due to the modulation will drive the signal up and down the resonant curve, producing simultaneous amplitude variations, which then can be detected using an envelope detector. Let us just take a case of FM and a specific tuned circuit and find the degree of AM.

Example 4.11

We have an FM signal centered at 10.7 MHz, with frequency deviation of 75 kHz. We have a purely parallel resonant circuit with a Q = 30, with resonant frequency such that 10.7 MHz is at the lower half-power frequency. Find the output voltage for $\Delta f = +75$ kHz and for -75 kHz.

SOLUTION   When we operate close to resonance, adequate accuracy is given by

$$V_o = \frac{V_i}{1 + j2Q\delta},$$

where δ is the fractional shift of frequency from resonance. If now, 10.7 MHz is the lower half-power point, we can say that $2Q\delta = 1$.

Hence, $\delta = 1/(2 \times 30) = (f_o - 10.7 \text{ MHz})/f_o$; $f_o = 10.881$ MHz.

Now, we evaluate the transfer function at 10.7 MHz ± 75.kHz.
We defined it as 0.7071 at 10.7 MHz. For 10.7 + 0.075 MHz, $\delta = (10.881 - 10.775)/10.881 = 9.774 \times 10^{-3}$, and the magnitude of the transfer function is $|1/(1 + j60\delta)| = 0.8626$.
Because the value was 0.7071 for the unmodulated wave, the modulation index in the positive direction would be

$$(0.8624 - 0.7071)/0.7071 = 0.2196 \text{ or } 21.96\%.$$

For (10.7 − 0.075) MHz, $\delta = (10.881 - 10.625)/10.881 = 0.02356$, and the magnitude of the transfer function is $|1/(1 + j60)| = 0.5775$. The modulation index in the negative direction is $(0.7071 - 0.5775)/0.7071 = 18.32\%$. So, modulation index is not the same for positive as for negative indices. The consequence of such asymmetry is that this process will be subject to harmonic distortion, which is why this process is not quite respectable.

EXERCISES 4.11
1. Repeat the example above if the Q of the circuit is 50.

ANSWERS   Amplitude modulation index is 35.22% up, 28.1% down, so it looks as though we have even worse curvature and harmonic distortion.

2. *Repeat the example above except that one places 10.7 MHz at the −6.02 dB point*

ANSWERS   Amplitude modulation index is 20.56% up, 15.34% down.

4.12   Lower Distortion FM Detection

We will assume that the reader has been left wanting an FM detector that has much better performance than the slope detector. A number of more complex circuits have a much lower distortion level than the slope detector. One, called the Balanced FM Discriminator, is shown in Figure 4.9.

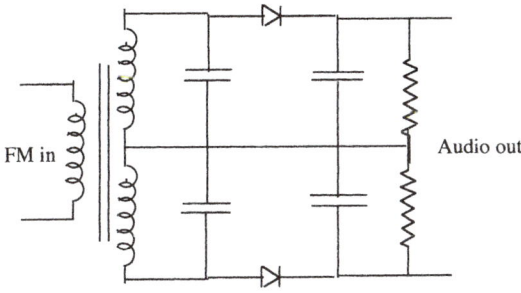

FIGURE 4.9
Balanced FM discriminator.

Basically, we may consider that the circuit contains two "stagger-tuned" resonant circuits, i.e., they are tuned equidistant on opposite sides of the center frequency, connected back to back. The result is that the nonlinearity of the resonant circuits balance each other out, and the FM detection can be very linear. The engineer designing an FM receiving system has a relatively easy job to access such performance; all that he/she must do is to spend the money to obtain high-quality components.

4.12.1   Phase-Locked Loop

The phase-locked loop is an assembly of circuits or systems that perform a number of functions to accomplish several operations, any one or more of the latter, perhaps being useful and to be capitalized upon. If one looks at a simple block diagram, one will see something like Figure 4.10

Thus, one function which will always be found is called a "voltage-controlled oscillator;" the linking of these words means that there is an oscillator which would run freely at some frequency, but that if a non-zero dc voltage is fed into a certain input, the frequency of oscillation will shift to one determined by that

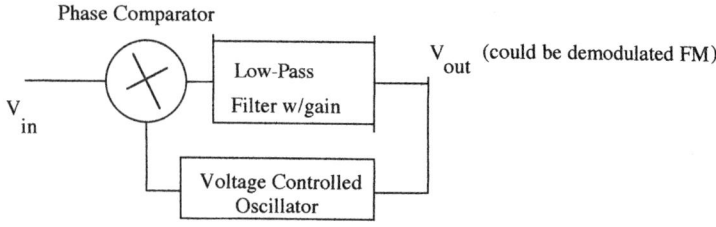

FIGURE 4.10
Basic phase-locked loop.

input voltage. Another function one will always find (although the nomenclature might vary somewhat) is "phase-comparison." The phase "comparator" will usually be followed by some kind of low-pass filter. Of course, if a comparator is to fulfill its function, it requires two inputs — the phases of which to compare. This operation might be accomplished in various ways; however, one method which might be understood from previous discussions is the analog multiplier. Suppose an analog multiplier receives the inputs $\cos \omega t$ and $\sin (\omega t + \phi)$; their product has a sine and a cosine. Now, a trigonometric identity involving these terms is

$$\sin A \cos B = 0.5\left[\sin(A + B) + \sin(A - B)\right].$$

Thus, the output of a perfect analog multiplier will be $0.5[\sin (2\omega + \phi) + \sin \phi]$. A low-pass filter following the phase comparator is easily arranged; therefore, one is left with a dc term, which, if it is fed to the VCO in such a polarity as to provide negative feedback, will "lock" the VCO to the frequency of the input signal with a fixed phase shift of 90°.

Phase-locked loops (abbreviated PLL) are used in a wide variety of applications. Many of the applications are demodulators of one sort or another, such as synchronous detectors for AM, basic FM, FM–stereo detectors, and in very precise oscillators known as "frequency synthesizers." One of the early uses seemed to be the detection of weak FM signals, where it can be shown that they extend the threshold of usable weak signals a bit.* This latter facet of their usefulness seems not to have made a large impact, but the other aspects of PLL usefulness are very commonly seen.

4.13   Digital Means of Modulation

The sections immediately preceding have been concerned with rather traditional analog methods of modulating a carrier. While the beginning engineer

---

* Taub, H. and Schilling, D.L. *Principles of Communication Circuits,* 2nd Edition, McGraw-Hill, New York, 1986, pp. 426-427.

can expect to do little or no design in analog communication systems, they serve as an introduction to the digital methods which most certainly will dominate the design work early in the twenty-first century. Certainly, analog signals will continue to be generated, such as speech, music, and video; however, engineers are finding it so convenient to do digital signal processing that many analog signals are digitized, processed in various performance-enhancing ways, and only restored to analog format shortly before they are fed to a speaker or picture tube. Digital signals can be transmitted in such a way as to use extremely noisy channels. Not long ago, the nightly news brought us video of the Martian landscape. The analog engineer would be appalled to know the number representing traditional signal-to-noise ratio for the Martian signal. The detection problem is greatly simplified because the digital receiver does not need at each instant to try to represent which of an infinite number of possible analog levels is correct; it simply asks, was the signal sent a one or a zero? *That* is simplicity.

Several methods of digital modulation might be considered extreme examples of some kind of analog modulation. Recall amplitude modulation. The digital rendering of AM is called "amplitude shift keying," abbreviated ASK.

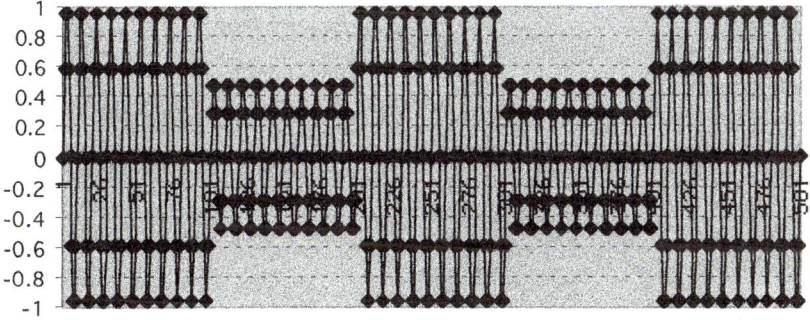

FIGURE 4.11
ASK (amplitude shift keying).

What this might look like on an oscilloscope screen is shown in Figure 4.11. For example, we might say that the larger amplitude signals represent the logic ones and smaller amplitudes represent logic zeroes. Thus, we have illustrated the modulation of the data stream 10101. If the intensity of modulation were carried to the 100% level, the signal would disappear completely during the intervals corresponding to zeroes. The 100% modulation case is sometimes called on–off keying and abbreviated OOK. The latter case has one advantage if this signal were nearly obscured by large amounts of noise; it is easiest for the digital receiver to distinguish between ones and zeroes if the difference between them is maximized. That is, however, only one aspect of the detection problem. It is also often necessary to know the timing of the bits, and for this one may use the signal to synchronize the oscillator in a phase locked loop: if, for 50% of the time, there is zero signal by which to be synchronized, the oscillator may drift significantly. In general, a format for digital modulation in which the signal may vanish utterly at intervals is to be adopted with caution and with full cognizance of one's sync problem. Actually, amplitude shift keying is not considered a very high performance means of digital signaling, in much the same was that AM is not greatly valued as a quality means of analog communication. What is mainly used is one or the other of the following methods.

4.13.1   Frequency Shift Keying

Frequency shift keying (abbreviated FSK) can be used in systems having very little to do with high data rate communications; for years it has been the method used in the simple modems one first used to communicate with remote computers. For binary systems, one just sent a pulse of one frequency for a logic one and a second frequency for a logic zero. If one was communicating in a noisy environment, the two signals would be orthogonal, which meant that the two frequencies used were separated by at least the data rate. Now, at first the modem signals were sent over telephone lines which were optimized for voice communications, and were rather limited for data communication. Suppose we consider that for ones we send a 1250 Hz pulse and for zeroes, we send 2250 Hz. In a noisy environment one ought not to attempt sending more than 1000 bits per second (note that 1000 Hz is the exact difference between the two frequencies being used for FSK signaling). Let us instead send at 250 bps. Twelve milliseconds of a 101 bit stream would look as in Figure 4.12.

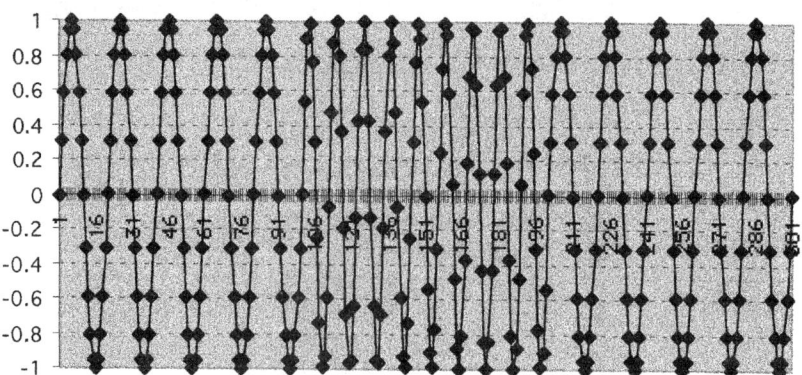

FIGURE 4.12
Frequency shift keying.

It is not too difficult to imagine a way to obtain FSK. Assuming one does have access to a VCO, one simply feeds it two different voltage levels for ones and for zeroes. The VCO output is the required output.

4.13.2 Phase Shift Keying

Probably the most commonly used type of digital modulation is some form of phase shift keying. One might simply say there is a carrier frequency f_c and that logic zeroes will be represented by $-\sin 2\pi f_c t$, logic ones by $+\sin 2\pi f_c t$. If the bit rate is 40% of the carrier frequency, the data stream 1010101010 might look as in Figure 4.13.

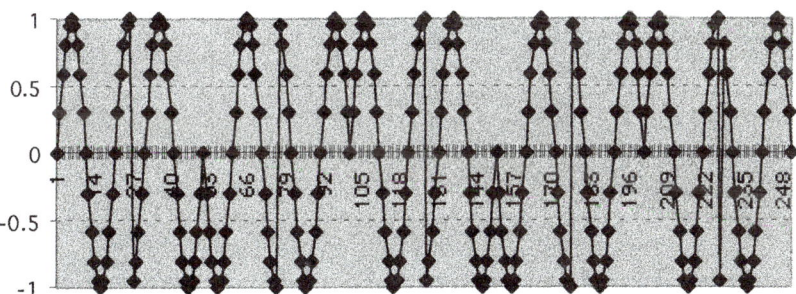

FIGURE 4.13
Phase shift keying.

In principle, producing binary phase shift keying ought to be fairly straightforward, if one has the polar NRZ (nonreturn to zero, meaning a logic one could be a constant positive voltage for the duration of the bit, zero being an equal negative voltage) bit stream. If then, the bit stream and a carrier signal are fed into an analog multiplier, the output of the multiplier could indeed be considered $\pm\cos \omega_c t$, and the modulation is achieved.

4.14 Correlation Detection

Many years ago, the communications theorists came up with the idea that if one could build a "matched filter," that is, a special filter designed with the bit waveform in mind, one would startlingly increase the signal-to-noise ratio of the detected signal. Before long, a practically minded communications person had the bright idea that a correlation detector would do the job, at least for rectangular bits. For some reason, as one explains this circuit, one postulates two signals, $s_1(t)$ and $s_2(t)$, which represent, respectively, the signals sent for logic ones and zeroes. The basics of the correlation detector are shown in Figure 4.14.

Now, a key consideration in the operation of the correlation detector is bit synchronization. It is crucial that the signal $s_1(t)$ be lined up perfectly with the bits being received. Then, the top multiplier "sees" $\sin \omega_c t$ coming in one input, and $\pm\sin \omega_c t + \text{noise}$ coming in the other, depending upon whether a

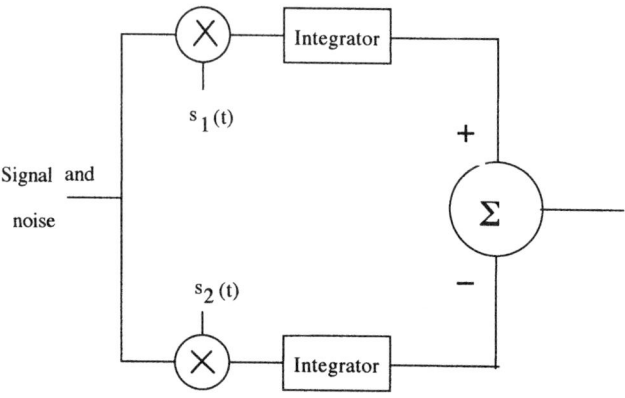

FIGURE 4.14
Correlation detector.

one or a zero is being received. If it happens that a one is being received, the multiplier is asked to multiply sin $\omega_c t$(sin $\omega_c t$ + noise). Of course,

$$\sin^2 \omega_c(t) = (1/2)(1 + \cos 2\omega_c t).$$

In the integrator, this is integrated over one bit duration, giving a quantity said to be the energy of one bit. The integrator might also be considered to have been asked to integrate $n(t)$ sin $\omega_c t$, where n is the noise signal. However, the nature of noise is that there is no net area under the curve of its waveform, so considering integration to be a summation, the noise output out of the integrator would simply be the last instantaneous value of the noise voltage at the end of a bit duration, whereas the signal output was bit energy, if the bit synchronization is guaranteed. Meanwhile, the output of the bottom multiplier was the *negative* of the bit energy, so with the signs shown, the output of the summing amplifier is twice the bit energy. Similar reasoning leads to the conclusion that if the instantaneous signal being received were a zero, the summed output would be *minus* twice the bit energy. It takes a rather substantial bit of theory to show that the noise output from the summer is *noise spectral density*. The result may be summarized that the correlation detector can "pull a very noisy signal out of the mud." And, we should assert at this point that the correlation detector can perform wonders for any one of the methods of digital modulation mentioned up to this point.

4.15 Digital QAM

Once the engineer has produced carrier signals that are 90° out of phase with each other, there is no intrinsic specification that the modulation must be

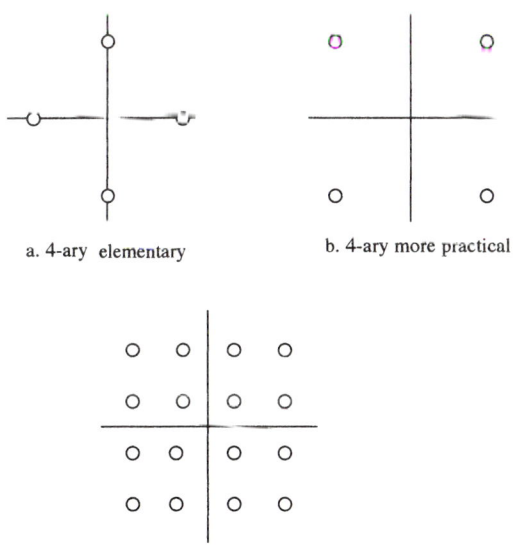

a. 4-ary elementary   b. 4-ary more practical

c. 16-ary constellation

FIGURE 4.15
Constellation showing carrier amplitudes and phase for M'ary signals.

analog, as is done for color TV. As a start toward extending the capabilities of PSK, one might consider that one sends bursts of several periods of $\pm\cos\omega_c t$ or $\pm\sin\omega_c t$. This is sometimes called "4-ary" transmission, meaning that there are four different possibilities of what might be sent. Thus, whichever of the possibilities is sent, it may be considered to contain two bits of information. It is a method by which more information may be sent without demanding any more bandwidth, because the duration of the symbol being sent may be no longer or shorter than it was when one was doing binary signaling, sending, for example, simply $\pm\cos\omega_c t$. This idea is sometimes represented in a "constellation," which, for the case we just introduced, would look like part a of Figure 4.15. However, what is more often done is as shown in Figure 4.15b, where it could be said that one is sending $\pm\cos(\omega_c t + 45°)$ or $\pm\cos(\omega_c t + 135°)$. It seems as though this may be easier to implement than the case of part a; however, the latter scheme lends itself well to sending 4 bits in a single symbol, as in Figure 4.15c.

Strictly speaking, one might consider "a" to be the constellation for 4-ary PSK. This leads also to the implication that one could draw a circle with "stars" spaced 45° apart on it and one would have the constellation for 8-ary PSK. The perceptive or well-informed reader might have the strong suspicion that crowding more points on the circle makes it possible to have more errors in distinguishing one symbol from adjacent ones, and would be correct in this suspicion. Hence, M'ary communication probably more commonly uses "b" or "c," which should be considered forms of QAM.

We will consider some of the performance aspects of digital modulation in the following chapter, when we consider their noise rejection properties.

5

Thermal Noise and Amplifier Noise Figure

5.1   Thermal Noise

Long before the modern-day engineer was born, Nyquist had thought of the
electrons in a conductor as behaving like a gas and had applied his understanding of thermodynamics to predict the nature of random voltages generated from the electrons jiggling around. Subsequently, van der Ziel added a
term that becomes important at extremely high frequencies, such as those of
visible light. The result is expressed most usefully in terms of a power spectral density which might be observed by instruments having infinite input
impedance but finite bandwidth, connected across resistance R. The result is
often written as

$$P(f) = 2R\left(\frac{h|f|}{2} + \frac{h|f|}{e^{h|f|/kT} - 1}\right),$$

where $|f|$ assumes that one will be integrating over positive and negative fre-
quencies with equal contributions and the "2" out front assumes such integration. Since the integrand *is* an even function, it is preferred to just multiply
by another 2 and integrate over positive frequency. The letter "h" is Planck's
constant,

$$6.2 \times 10^{-34} \text{ J - sec,}$$

with the result that the first term in the bracket is negligible until one gets
to very high frequency, such as visible or ultraviolet light. The letter "JK" is
Boltzmann's gas constant,

$$1.38 \times 10^{-23} \text{ K,}$$

K stands for Kelvin degrees, and T is Kelvin temperature, which the reader
may recall is Celsius temperature + 273. Now, at low frequencies, further

approximations are possible (and very welcome). It may be recalled that one can expand the exponential function in an infinite series, but that one can keep just two terms if the exponent is small compared with one. Suppose we decide that "small compared to one" means one tenth or less. Let's assume temperatures near a typical (wintertime) room temperature of 290K, and find what maximum frequency this implies.

$$\frac{h|k|}{kT} = 0.1; \quad |f| = \frac{0.1 \times 1.38 \times 10^{-23} \times 290}{6.2 \times 10^{-34}} = 645 \text{ GHz}$$

One will never approach such frequencies except in the optics lab, so for most frequencies with which we would want to do noise calculations, we *may* write e^x as $1 + x$, to a good approximation. Hence, we change the spectral density expression to

$$P(f) = \frac{4Rh|f|}{1 + \dfrac{h|f|}{kT} - 1} = \frac{4Rh|f|}{\dfrac{h|f|}{kT}} = 4kTR.$$

Thus, we see that at low frequencies, the actual frequency is removed from the power spectral density expression, and it is a constant independent of frequency. In the early days when physicists were learning about light, they noted from passing sunlight through a prism that light, which is apparently white, is actually made up of a spectrum of colors. Similarly, at low frequencies, thermal noise has the same amount of noise power in each Hertz of bandwidth, so thermal noise is said to be "white noise." In Figure 5.1 we have plotted the thermal noise power density vs. powers of ten in frequency to dramatize this behavior, at sub-optical frequencies.

The result is often stated that if one connected a voltmeter of infinite input impedance, but a sharp cut-off at a bandwidth B, one would observe

$$v^2 = 4kTRB.$$

Example 5.1
Find the thermal noise voltage that one should observe across a 250k resistor at room temperature, using a voltmeter with bandwidth of 1 MHz.

SOLUTION $\quad v = \sqrt{4 \times 1.38 \times 10^{-23} \times 290 \times 2.5 \times 10^5 \times 10^6} = 63.3 \text{ } \mu v.$

Before the student decides to verify this number experimentally, it should be quickly mentioned that, in the typical lab or industrial setting, many other factors besides thermal effects are generating interfering voltages, and that if one is determined to try the experiment, one should be sure to try it in a

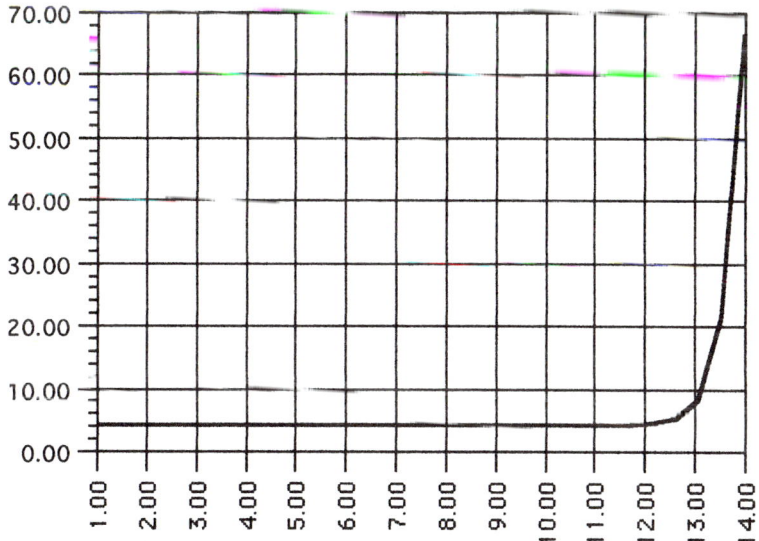

FIGURE 5.1
Thermal noise power spectral density vs. powers of 10 in frequency.

"screen room," where most interfering voltages, except, thermally generated ones, are screened out.

5.1.1 "Available" Noise Power

In circuit theory, the word "available" means the amount of power that will be delivered to a load which is impedance-matched to the source. This means, if we could connect a noiseless resistor equal to R across the noisy R, what power would be observed? First, one should realize that one has constructed a 2:1 voltage divider, so the voltage across the lossless resistor would be obtained from

$$v^2 = \left(\frac{1}{2}\right)^2 \times 4kTRB = kTRB.$$

If one wants to know available noise *power*, one divides by R, getting $N_{avail} = kTB$.

Example 5.1.1

1. *One can generally expect that laboratory signal generators have their output impedances at room temperature. What available noise power into a matched load can one expect from a generator with 1 MHz bandwidth?*

SOLUTION $kTB = 1.38 \times 10^{-23} \times 290 \times 10^{6} = 4 \times 10^{-15}$ watts .

2. The main departures from room temperature of sources of noise come from antennas. In the AM band, source temperatures of $10^7 K$ are not unusual. In the FM band, one should expect source temperatures of about 1000K. However, UHF TV may experience source temperatures around 100 and around 2–4 GHz; directive antennas pointed straight up in the air may see temperatures of 10 or 20K. This has very happy implications for satellite TV. Let us find available noise power in the AM band where bandwidth is 10^4.

SOLUTION $kTB = 1.38 \times 10^{-23} \times 10^7 \times 10^4 = 1.38 \times 10^{-12}$ watt, or 1.38 pw.

EXERCISES 5.1
1. Find the available noise power into an FM receiver with bandwidth of 200 kHz, matched to the antenna which sees a "sky temperature" of 2000K.

ANSWER   5.52 femtowatts $= 5.52 \times 10^{-15}$ watts.
2. What is the available noise power into a satellite TV receiver with bandwidth of 12 MHz in the antenna is pointed in a direction in the sky where the temperature is 50K?

ANSWER   8.28 fw.

5.2   Amplifier Noise Figure and Excess Temperature

In Candide's "best of all possible worlds," the noise out of an amplifier would be simply the power gain of the amplifier times the noise being fed into the amplifier. In this real world, the output noise is always a little higher. This leads to the definition of a quality factor for an amplifier which we will call its "noise factor," which is defined at standard room temperature of 290K as the ratio of the amplifier's input signal-to-noise ratio divided by its output signal-to-noise ratio. This is given by the symbol F and may be written

$$F = \frac{(S/N)_{in}}{(S/N)_{out}} = \frac{S_{in}/kT_oB}{GS_{in}/(GkT_oB + N_{excess})} = \frac{(GkT_oB + N_{excess})}{GkT_oB}$$

It should be noted that noise factor (and its related quality factor "noise *figure*") are only useful concepts if the source driving the amplifier is at the so-called room temperature $T_o = 290K$. Otherwise, an alternate concept, the "excess temperature" of the amplifier is very useful. It is as though implied that excess noise power results from higher than expected temperature at the input, and one can write

$$N_{excess} = GkT_{excess}B;$$

hence, one can write $F = 1 + T_e/T_o$. Indeed, there will be times when one thinks the only use of noise factor is that it enables one to find the excess temperature. Note also that the noise factor is always >1.

Example 5.2

1. Find the noise factor corresponding to an excess temperature of 58K.

SOLUTION $F = 1 + \dfrac{58}{290} = 1.2.$

Originally, the IEEE thought that if one took $10 \log_{10} F$, it would be known as "noise figure." In some texts, it receives the symbol F_{dB}.

2. Find the noise figure corresponding to an excess temperature of 58K.

SOLUTION $F_{dB} = 10 \log_{10}(1.2) = 0.79$ dB.

EXERCISES 5.2

1. Find noise factors and noise figures corresponding to excess temperatures of 100K, 150K, 290K, and 1000K.

ANSWERS

$$10 \log_{10}(1.344) = 1.29 \text{ dB}$$

$$10 \log_{10}(1.517) = 1.81 \text{ dB}$$

$$10 \log_{10}(2) = 3.01 \text{ dB}$$

$$10 \log_{10}(4.45) = 6.46 \text{ dB}$$

5.3 Noise Temperature and Noise Factor of Cascaded Amplifiers

In Figure 5.2, we show an incomplete schematic of a cascade of stages of amplification with differing amounts of gain and the different amounts of excess noise which they add. It must be noted that excess temperature is related *only to the input of each amplifier stage*. Therefore, the noise after two stages of amplification is

$$N_{out_2} = G_1 G_2 kB\left(T_s + T_{e_1}\right) + G_2 kBT_{e_2},$$

and after the third stage,

$$N_{out_3} = G_3\left[G_1 G_2 kB\left(T_s + T_{e_1}\right) + G_2 kBT_{e_2}\right] + G_3 kBT_{e_3}.$$

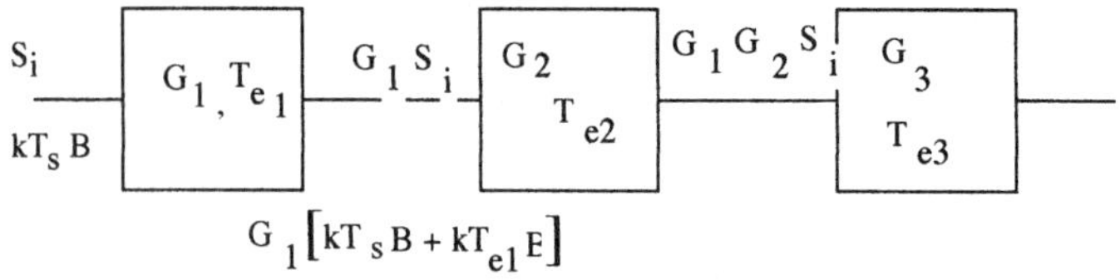

FIGURE 5.2
Signal and apparent noise present between stages of amplifier cascade.

Normally, the requirements in communication systems would be in terms of output signal-to-noise ratio. In terms of the relations here, we have obtained

$$\left(\frac{S}{N}\right)_{out} = \left(\frac{G_1 G_2 G_3 S_i}{G_3\left[G_1 G_2 kB\left(T_s + T_{e_1}\right) + G_2 kBT_{e_2}\right] + G_3 kBT_{e_3}}\right) =$$

$$\left(\frac{S_i}{kT_s B}\right)\frac{1}{1 + T_{e_1}/T_s + T_{e_2}/G_1 T_s + T_{e_3}/G_1 G_2 T_s}.$$

The first term in the last relation is the input signal-to-noise ratio. Being added in the denominator, we can see how temperatures of the various stages work to reduce the output signal-to-noise ratio. Of special interest is the way that gains in the first stages tend to reduce the effects of later-stage excess temperature. This is not the form of equation usually seen for cascades, but it is *always correct, regardless of source temperature* T_s. If, and only if, it should happen that the source is at standard temperature, we can write the noise factor for the cascade shown

$$F = F_1 + \frac{F_2 - 1}{G_1} + \frac{F_3 - 1}{G_1 G_2}$$

If more stages need to be accounted for, one can continue the pattern with each successive term being the next noise factor minus 1 and divided by one more stage gain. In the expression in terms of temperatures, each successive excess temperature gets divided by one more gain.

Example 5.3
Suppose the first stage has power gain of 50 and excess temperature of 87K, the second stage has gain of 100 and temperature of 290K, and the third one has excess temperature of 580K. How much will input signal-to-noise ratio be reduced? Assume the source is at standard temperature $T_0 = 290$K.

SOLUTION Taking just the second traction in the excess temperature equation, we have

$$\cfrac{1}{1+\cfrac{87}{290}+\cfrac{290}{50\times290}+\cfrac{580}{50\times100\times290}} = \cfrac{1}{1+0.30+0.02+0.0004} = 0.757,$$

which is the amount by which input signal-to-noise ratio is reduced.

EXERCISES 5.3

This exercise will demonstrate how very important it is to have good gain in the early stages of an amplifier for good noise performance. Suppose, in the amplifier above, the only change is that the first stage power gain is only 5. Now find the reduction in output signal-to-noise ratio.

ANSWER 0.649.

5.3.1 Noise Factor of an Attenuating Cable

As was demonstrated in Chapter 1, transmission lines such as coaxial cable all have a "characteristic impedance" such that if that impedance is used as load impedance, the input impedance at the other end will also be equal to the characteristic impedance, and hence the line is matched at both ends, if we let it be driven by a generator also having the characteristic impedance. The result is that the load is correct to obtain available noise power whether it is connected as load or connected directly to the generator; it gets the available noise power at either place. The cable may be considered to be at room temperature, so we can say its noise figure is input S/N divided by output S/N. So, we have

$$F = \frac{(S/N)_{in}}{(S/N)_{out}} = \frac{S_{in}}{S_{out}},$$

because the noise powers are the same. However, we have

$$S_{out} = KS_{in},$$

where K is a number less than one, and the noise factor is $1/K$. Here then can be a double dose of bad news, because this cable can be the first "amplifier stage," a high frequency signal passes through, and it not only has a noise figure to worry about, but it also has a gain less than one to divide into other terms. We will see this problem in examples we work.

Example 5.3.1

A high frequency receiver having a noise figure of 4 dB is driven by cable having 3.01 dB of attenuation. What is the resulting noise factor and figure of the cascade?

SOLUTION   We must first convert all noise figures into noise *factors*.

$$-3.01 = 10\log_{10} F;$$

$$F = 10^{-3.01/10} = 1/2.$$

This means that the power gain of the first amplifier in the chain is $1/2$.

$$F_2 = 10^{4/10} = 2.51, \text{ so overall,}$$

$$F = 2 + \frac{2.51 - 1}{0.5} = 5.02, \text{ and}$$

$$F_{dB} = 10\log_{10} 5.02 = 7.007 \text{ dB.}$$

EXERCISES 5.3

1. Rework the example above if the cable has only 1.0 dB of attenuation.

ANSWER   5.0 dB. So, we see, small amounts of attenuation have less drastic effects.

2. Now, consider a good means of mitigating the effects of attenuation. Suppose that we precede the cable by a low noise preamplifier having power gain of 10 and noise factor of 2.0. Find the overall noise factor and figure for 1 dB and 3.01 dB of loss in the cable, which is now of course amplifier number 2.

ANSWERS $F = 2.216$ or 3.46 dB for 1 dB cable;
$\qquad\qquad\quad F = 2.40$ or 3.80 dB for 3.01 dB cable.

3. In a cascade of amplifying stages, the first has a noise figure of 1 dB and a power gain of 10 dB, the second stage noise figure is 3.01 dB and power gain is 20 dB, and the third has noise figure of 6.02 dB and gain of 23 dB.

   a. What is the noise figure of the cascade?

ANSWER   1.34 dB.

   b. What is the excess noise temperature of the cascade?

ANSWER   99K.

   c. If this amplifier is fed by a source at room temperature, what must be the input signal-to-noise ratio for an output ratio of 10? (Hint: Remember that the definition of noise factor is $(S/N_{in})/(S/N_{out})$ when the source is at room temperature.)

ANSWER   13.62

    d. The input signal is 10^{-14} watts, or 10 femtowatts. The antenna is directed at a point in the sky where noise temperature is 30K. The amplifier bandwidth is 500 kHz. Find the input S/N and the ratio at the output of the amplifier.

ANSWER   48.3, 11.2.

4. *The amplifier 3 is fed through a cable having a loss of 0.5 dB. Now rework all parts of 3, with the added stage up front.*

ANSWERS   4.23 dB, 478.5K, 26.49, 48.3, and 2.85.

5. *Again, we will fight the cable loss problem by preceding it with a preamplifier having noise figure of 1 dB and gain of 10 dB. Find the noise figure of the newest cascade, its noise temperature, and what must be the input signal for output S/N of 10 if the source temperature is still 30K.*

ANSWERS   1.178 dB, 90.4K, and 8.3 fw.

5.4   Transistor Gain and Noise Circles

When one gets to the actual point of low-noise amplifier design, it is often found that the operating point for maximum gain may be very different from that for minimum noise figure. However, there exists a technique for finding circles upon which the gain or the noise figure is a constant, which may be plotted on a Smith chart. Then one can choose an operating point so as to make a good compromise between the gain of the stage and its noise performance. Let us again consider the MRF571 transistor. In Chapter 1.9, we found the input reflection coefficient for simultaneous conjugate match as $\Gamma_{GM} = 0.890\angle{-178.71°}$. However, the data sheet for the transistor says that at this collector current $I_c = 5.0$ mA, and at a frequency of 1.0 GHz, the minimum noise figure of 1.5 dB is obtained for a reflection coefficient connected to the input of $\Gamma_G = 0.48\angle{148°}$. The sheet also gives a resistance needed to compute the noise circles, $R_n = 7.5\Omega$. The procedure is that for a succession of noise figures higher than the optimum, one computes an intermediate variable N, where

$$N = \frac{F - F_{min}}{4R_n/Z_o}\left|1 + \Gamma_{opt}\right|^2.$$

Then the noise circles are centered at

$$C_{N} = \Gamma_{opt}/(1+N), \text{ and have radii } R_N - \frac{\sqrt{N\left(N+1-\left|\Gamma_{opt}\right|^2\right)}}{1+N}.$$

Let us tabulate all of this for noise figures going up from F_{opt} by tenths of a dB.

F_{dB}	F	N	$|\Gamma_{opt}|/(1 + N)$	R_N
1.6 dB	1.445	0.0309	0.466	0.154
1.7	1.479	0.0625	0.452	0.215
1.8	1.514	0.0949	0.438	0.262
1.9	1.549	0.128	0.426	0.300
2.0	1.585	0.162	0.413	0.334

As for gain, if one uses Γ_{GM} at the input, the available gain is 14.0 dB. Let us find circles for increments of minus 0.5 dB in gain; one calculates a gain factor called g_p which is an actual gain divided by the square of the magnitude of S_{21}. Then, the centers of gain circles will be at

$$C_p = \frac{g_p C_1^*}{1 + g_p\left(|S_{11}|^2 - |\Delta|^2\right)}.$$

The radii of the circles will be

$$r_p = \frac{\sqrt{1 - 2K|S_{21}S_{12}|g_p + |S_{21}S_{12}|^2 g_p^2}}{1 + g_p\left(|S_{11}|^2 - |\Delta|^2\right)}.$$

Noting that the denominator is the same in both expressions, we tabulate it as part of our systematic progress towards these circles.

dB Gain	g_p	Denom	Center	Radius
13.5	2.487	1.900	0.834	0.127
13.0	2.217	1.802	0.785	0.190
12.5	1.976	1.715	0.734	0.246
12.0	1.761	1.637	0.686	0.299
11.5	1.569	1.568	0.638	0.350

These circles have been graphed onto a Smith chart as shown in Figure 5.3

These circles demonstrate that one will always need to make compromises between gain and noise figure. If one would demand the lowest noise figure of 1.5 dB, one has to settle for less than the minimum gain graphed, or roughly 11.3 dB. For the lowest noise figure graphed, the best gain available is 13.0 dB. Two circles which just touch each other are for a noise figure of 1.7 dB and gain of 12.5 dB. How, the reader may ask, does one make a choice among the various possibilities? The answer is that one must look for the best overall performance based upon the noise factor of the total *following* stages. The following example illustrates how this might be done.

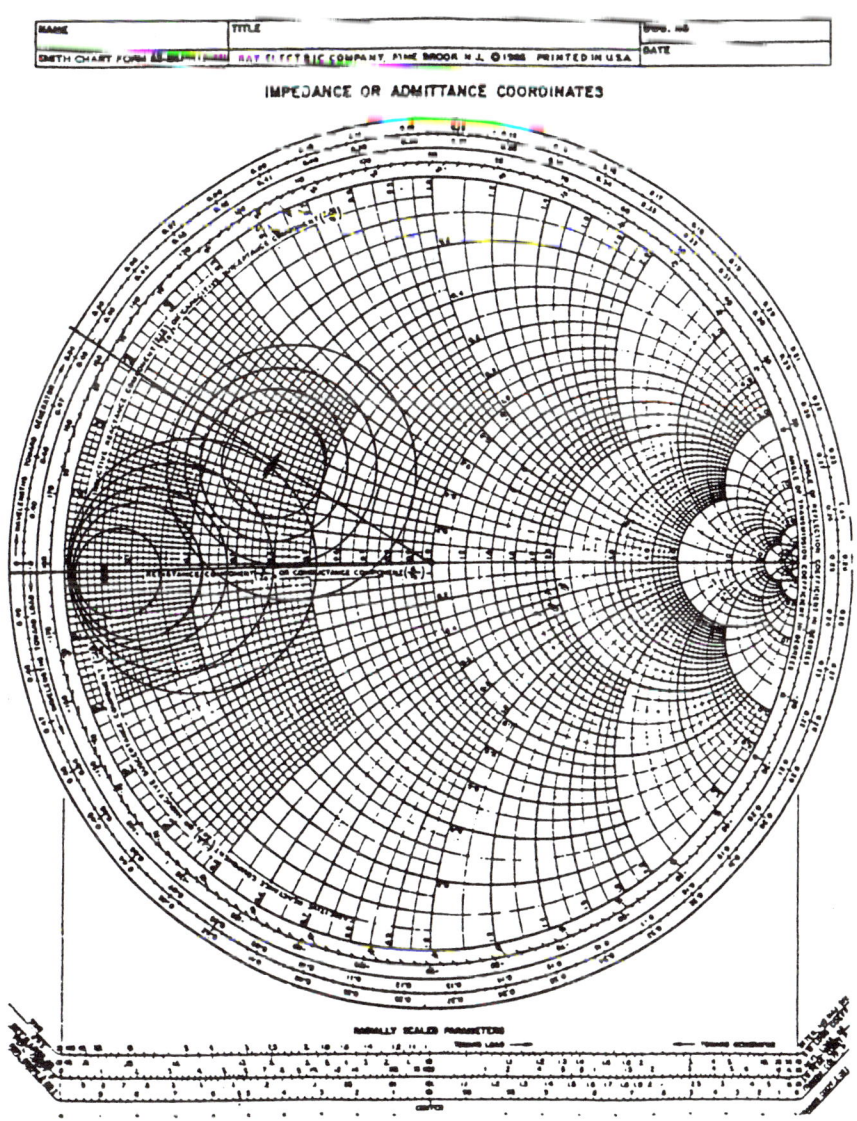

FIGURE 5.3
Gain and noise circles for Example 5.4.

Example 5.4

Assume that the overall noise figure of the following stages is 3.0 dB. Let us consider the case of F_1 of 1.5 dB and the gain of the first stage is 11.3 dB. What is the resulting overall noise figure?

SOLUTION $F_1 = 10^{0.15} = 1.4125$
$$G_1 = 10^{1.13} = 13.489$$

$$F_2 = 10^{0.300} = 1.995$$
$$F = 1.4125 + 0.995 / 13.489 = 1.486$$

EXERCISES 5.4

1. *In order to really see the circumstances when it is important to emphasize first stage gain and when it is preferable to have minimum first stage noise factor, the reader is asked to compute the overall noise figure for the three operating points mentioned, for several values of the following noise factors. For convenience in spelling out the problem, the following tabulation is given as a start and the reader is directed to try for the answers in the latter three columns:*

ANSWERS

G_1	F_1	F for $F_2 \rightarrow 3.0$ dB	5.0 dB	7.0 dB
11.3 dB→13.489	1.5 dB→1.4125	1.486	1.573	1.710
12.5 dB	1.7 dB	1.535	1.600	1.705
13.0 dB	2.0 dB	1.598	1.656	1.745

Generalizing these results, we see that when the later stages have a low noise factor, best overall noise factor happens with the lowest value for the first stage. However, as one gets higher noise factors for the later stages, it becomes preferable to have higher gain in the first stage even though its noise factor is compromised.

2. *Considering the widespread nature of noise problems, it is a little surprising how scarce the data are for drawing noise circles. We will borrow some data from one of the earliest application notes published by Hewlett-Packard. This was perhaps the first widespread promotion of the application of scattering parameters to optimize design and assess the results for high frequency transistors.*
The data, at 4 GHz, were: $S_{11} = 0.552\angle169°$
$$S_{12} = 0.049\angle23°$$
$$S_{21} = 1.681\angle26°$$
$$S_{22} = 0.839\angle{-67°}$$
$$F_{min} = 2.5 \text{ dB}$$
$$\Gamma_{opt} = 0.475\angle166°$$
$$R_o = 3.5 \ \Omega$$

ANSWERS $|\Delta| = 0.419$, $K = 1.012$, so transistor is unconditionally stable at this frequency. Figure 5.4 is a copy of the author's solutions for the noise and gain circles for the transistor.

5.5   Narrowband Noise

It was previously established for most of the frequency range presently used for communications that the noise one must deal with is "white," which means that the density of the power in terms of bandwidth is *uniform*. Indeed,

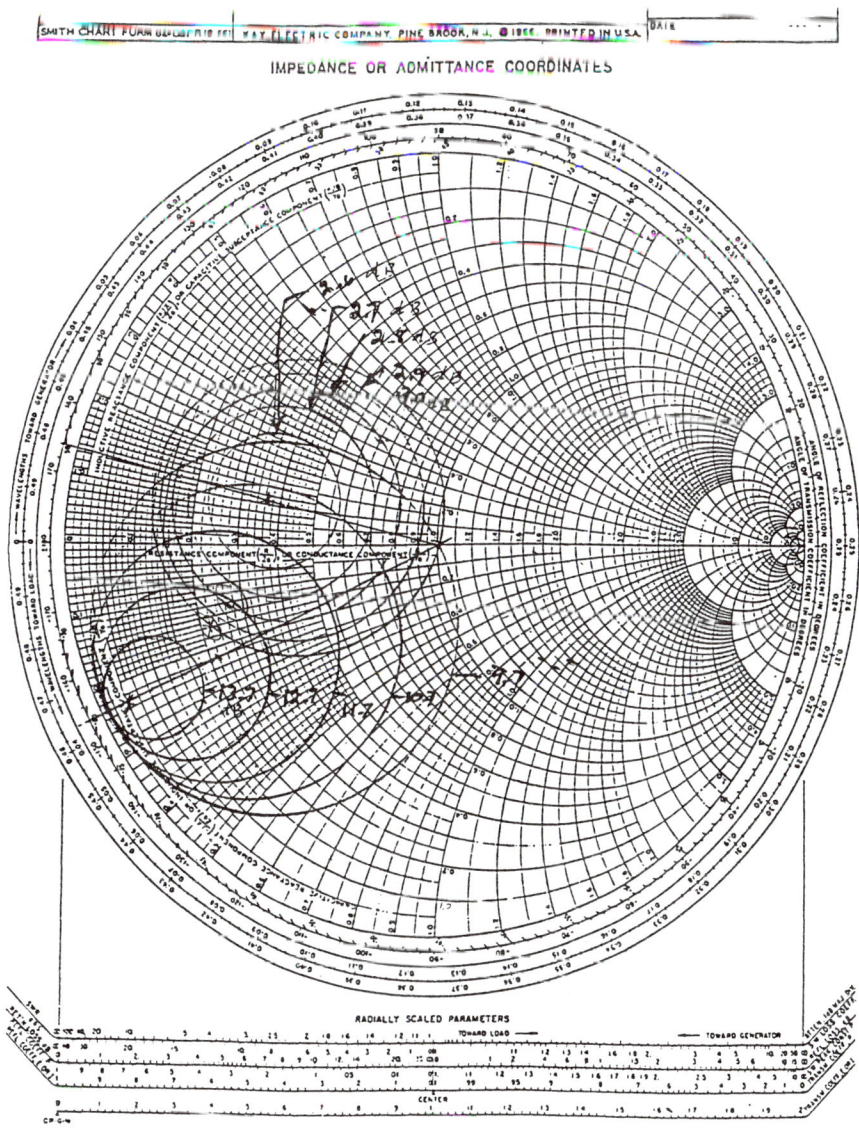

FIGURE 5.4
Gain and noise circles for Exercise 5.4.2.

if we accept that the available noise power is kTB, where T is the system noise temperature and B is now the "effective noise bandwidth," we can then quote the noise power spectral density as kT. In order to understand the effects of high frequency noise upon the output noise of a receiver, it may well be convenient to consider the noise power in a narrow band as being that of a sinusoid of variable amplitude and phase, having the frequency of the middle of the band. If one doubts the experimental validity of such an assumption, one is invited to feed the voltage from a white noise generator into a low-pass filter and to feed the filter output to a locked-in oscilloscope. What one sees is apt to contain *very stable zero-crossings with spacings related to the filter bandpass*, albeit a signal which bounces around in amplitude greatly, and thus may be interpreted as varying in amplitude and phase. Let us examine the implications of this in a sort of discussion/example.

Suppose we have an unmodulated carrier at 1 MHz and a passband 10 kHz wide, centered at 1 MHz. Say we want to split the passband into 10 equal smaller bands. The bands would be 1000 Hz wide. Hence, as shown in Figure 5.5,

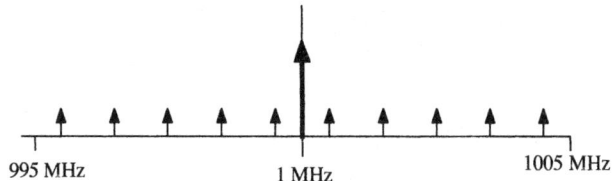

995 MHz   1 MHz   1005 MHz

FIGURE 5.5
Carrier plus phasor representation of narrowband noise.

our approximation would be to show equal noise spectral entries centered at 995.5, 996.5, 997.5, 998.5, 999.5, 1000.5, 1001.5, 1002.5, 1003.5, and 1004.5 kHz. Now, the fact that both amplitude and phase are varying may be interpreted as saying that we have, for example, varying amplitudes of *both*

$$\cos\left(2\pi \times 1.0005 \times 10^6 t\right) \text{ and } \sin\left(2\pi \times 1.0005 \times 10^6 t\right).$$

Consider the effects of the noise on the modulated envelope of the carrier. We may expand the cosine term as

$$\cos\left(2\pi \times 10^6 t\right)\cos 2\pi \times 500\ t - \sin\left(2\pi \times 10^6 t\right)\sin 2\pi \times 500\ t.$$

Now, let us add to this a large carrier term, for example, 100 cos 2π x 106t. The resulting envelope is the square root of the sum of the squares of the coefficients of the two carrier frequency terms, specifically,

$$E(t) = \sqrt{100 + \cos(1000\pi t)^2 + \sin^2(1000\pi t)}$$

$$= \sqrt{10^4 + 200\cos(1000\pi t) + \cos^2(1000\pi t) + \sin^2(1000\pi t)}\ .$$

Now, if we add cosine squared and sine squared of anything, we get unity every time. However, in our present situation, that adds only 0.01% to the

10,000 up front, and may be ignored. If we then factor the 10,000 out of the radical, what we have left is

$$E(t) \cong 100\sqrt{1 + 0.02\cos(1000\pi t)}.$$

Applying the binomial theorem, we say the square root of 1 + pretty small is approximately 1+ (pretty small)/2, so the effect on the envelope from that sideband is

$$E(t) = 100 + \cos(1000\pi t).$$

The noise sideband retained an identity; a sideband 500 Hz from the carrier affects the envelope as noise centered at 500 Hz, and would have that identity if the noise signal is passed through an envelope detector. The spectrum shown would contribute another 500 Hz term from the 999.5 kHz component, two 1500, 2500, 3500, and 4500 Hz components. Admittedly, our figure gives a very oversimplified view of the noise that is received. Perhaps the reader would feel more satisfied to picture a forest of noise components separated by 1 Hz.

In the discussion above, there is another oversimplification which must be dealt with. A principle related to the equipartition of energy in statistical mechanics decrees that in addition to the noise sidebands listed, there will also be *sines* of 1000($\pm$0.5, 1.5, etc.). Expanding the term corresponding to the example above and adding to the large carrier term, one has:

Example 5.5.1
The sideband may be expanded as

$$\sin(2\pi \times 10^6 t)\cos 1000\pi t + \cos(2\pi \times 10^6 t)\sin 1000\pi t.$$

The paired sideband *below* the carrier contributes the same thing except that the second term is negative and cancels the second term of the upper sideband. Now, the resulting envelope simply contains

$$E(t) = \sqrt{10^4 + 4(\sin^2(1000\pi t) + \cos^2(1000\pi t))} = \sqrt{10,004} \cong 100.02.$$

Stated in words, the sidebands containing the *sine* terms contribute nothing to the envelope. Hence, half the available noise power is ignored by the envelope detector.

There is an unfortunate circumstance with the envelope detector; there must be a strong carrier present for envelope detection to work at all. Carrier power does indeed count in determining whether the receiver is above threshold or not. However, *the presence of the strong carrier contributes nothing to the signal output.* This is a reason why it may be necessary to calculate modulation efficiency in AM systems. An output signal-to-noise ratio which was doubled by detection will be reduced proportional to modulation efficiency. Let us illustrate this with an example.

Example 5.5.2

An AM system with a modulation efficiency of 40% is right at threshold. What will be its signal-to-noise ratio after envelope detection?

SOLUTION   Threshold means S/N of 10. This is doubled to 20 by the detection process but reduced proportionally to the efficiency. The net result is detected S/N = $10 \times 2 \times 0.4 = 8.0$.

There is yet one more circumstance allowing one to increase S/N, even after detection. Suppose that ones receiver had more bandwidth than was needed to handle the signal. It may not be at all convenient to reduce the receiver bandwidth to the minimum value needed. There is still the possibility of using a low-pass filter on the detected signal. **Because the extra noise maintained its identity and spectrum,** it can be filtered out by the so-called post-detection filter, a low-pass filter which passes the signal intact but discards the higher frequency noise. This becomes very important in FM; best performance is found for $\beta \gg 1$. Carson then decrees that to handle this signal, one needs much more bandwidth than simply twice maximum modulation frequency. The Carson bandwidth must be used in the threshold calculation. However, the post-detection filter can deliver quite an acceptable output S/N in FM systems, even if the signal were exactly at threshold at the receiver input.

Further Example

Suppose that in the previous example we know that our receiver had 50% more bandwidth than was needed to pass the signal. What would be the signal-to-noise ratio if one used a post-detection filter with exactly the bandwidth needed to pass the signal?

SOLUTION   We simply divide out the receiver bandwidth and multiply in the bandwidth of the post-detection low-pass filter. Since noise is in the denominator here, we would have S/N = $12 \times 150\% / 100\% = 18$. It is important to emphasize that a noise-canceling effect similar to that in envelope detection occurs in synchronous detection. Remember that the synchronous detection process multiplies the high frequency signal by a carrier, which translates the modulated signal down to baseband and does a similar shift to the noise sidebands. This process again has terms canceling each other in the terms involving the sines. Thus, in synchronous detection, the input S/N is again simply doubled. However, if it happened that one was using more bandwidth than was necessary to pass the signal, a post-detection filter could again strip off the extra noise. Summarizing then, we may generalize the result after demodulation of amplitude modulated signals, whether the signal detection is done by synchronous detection or by envelope detection. Both methods successfully ignore 50% of the effective input noise power; therefore, the process of demodulation doubles the input signal-to-noise ratio, no matter what demodulation method is used. We may expect this result whenever the signal is over a threshold defined by received signal power being at least 10 times noise power. Note that if one is receiving too much noise power to satisfy threshold requirements, there is no repairing the situation if one is using an envelope detector. When the noise is too strong to satisfy threshold, it seizes control of the envelope so that a large amount of noise is detected. If, however, one has enough signal to satisfy threshold requirements and is using more bandwidth than the *signal* requires, one can further improve the signal-to-noise ratio by reducing bandwidth to that required for the signal.

EXERCISES 5.5

1. *In a synchronously detected DSB–SC system, the received signal-to-noise ratio is 50. However, the receiver has 60% more bandwidth than is needed. Find the output signal-to-noise ratio if an appropriate post-detection filter is used.*

ANSWER   160.

2. *An AM communication system having a modulation efficiency of 20% is received with a receiver input S/N of 40. However, the receiver has 160% more bandwidth than is needed. Find the output S/N if the appropriate post-detection filter is used.*

ANSWER   206.

5.6   Noise and Interference in FM Systems

It is interesting that the effects of noise in a FM system can be easily illustrated and understood in terms of an interfering signal which happens to fall in the passband of the FM receiver. Consider Figure 5.6.

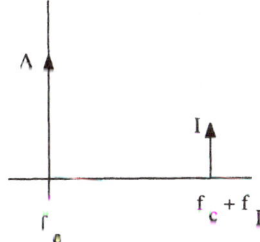

FIGURE 5.6
FM carrier plus interfering signal.

The specifications we place on what we have illustrated there is that the amplitude of the carrier A is strong compared to the amplitude I of the interfering signal. The frequency of the interfering signal is an amount f_I above the carrier. We take steps to combine the signals. We have

$$A \cos(\omega_c t) + I \cos(\omega_c + \omega_I)t.$$

Using the equation for cosine of a sum, we expand the interference term, and get

$$A \cos(\omega_c t) + I \cos(\omega_c t)\cos(\omega_I t) - I \sin(\omega_c t)\sin(\omega_I t).$$

If now we would combine the cos ($\omega_c t$) terms, we get A + I cos $\omega_I t$. However, our assumption that the carrier term is much larger than the interfering term permits us to neglect it in comparison with the carrier term. Now, let us look for a phase shift term; its tangent will be the negative coefficient of sin $\omega_c t$ over the coefficient of cos ($\omega_c t$), or approximately

$$\theta = \tan^{-1}(I \sin \omega_I t / A) \approx I \sin \omega_I t / A.$$

Now, what a frequency detector will detect is instantaneous frequency, which is the time derivative of phase; thus,

$$f_i = \frac{1}{2\pi}\frac{d\theta}{dt} = \frac{I}{A}\frac{2\pi f_I}{2\pi}\cos \omega_I t = \frac{I f_I}{A}\cos(2\pi f_I t).$$

Some interesting characteristics of FM interference are visible here; for one thing, the amplitude is inversely proportional to carrier amplitude, hence the stronger the carrier, the weaker the interfering signal. This is the source of "quieting," the phenomenon that reduces the effects of noise as signal strength increases. Another interesting fact is that for an interfering signal separated from the carrier by an amount f_I, the signal out of an FM detector will be *exactly at f_I*.

However, perhaps the most difficult circumstance is that *amplitude* of the interfering signal is also *proportional* to f_I! This can be a great disadvantage; the reader can perhaps remember that when narrowband frequency was studied in connection with AM, it retained its identity at f_I, but its amplitude was *independent of f_I*. Thus, while the noise into the FM receiver is "white," i.e., its amplitude was independent of frequency, the noise out of an FM detector has a "parabolic" spectral density, which is to say, noise power density depends upon the *square* of its separation from f_c. If not compensated for, this could make FM a rather noisy medium.

The first compensation for FM noise sounds crude indeed; a low-pass filter attenuates signals in inverse proportion to frequency at frequencies far above

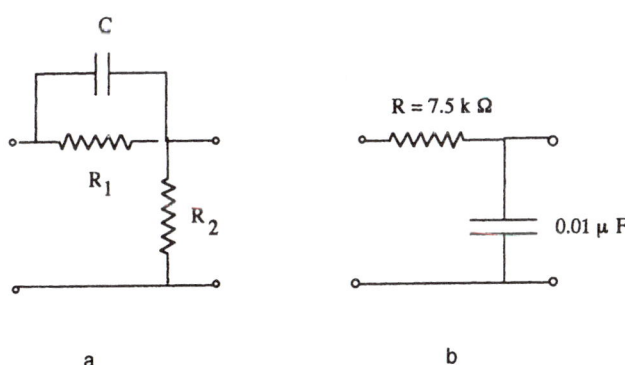

FIGURE 5.7
a. Pre-emphasis and b. de-emphasis circuits.

its corner frequency. Thus, if one places the corner frequency low in the audio range, it will eliminate large amounts of perceived noise. That is the good news; the bad news is that it will also remove so much signal as to make the claim of high fidelity for FM a bit ridiculous. However, some unnamed but clever engineer proposed that since one is going to really de-emphasize high frequencies after detection, they should be emphasized by an equal amount before they are transmitted. What was settled upon for FM standards was that the low-pass filter should have a time constant of 75 μsec, which corresponds to a corner frequency of 2122 Hz. But before the modulating signal went to the modulator, it was to be fed into a so-called pre-emphasis circuit. This circuit is as shown in Figure 5.7a. The low-pass filter we first spoke of is shown in Figure 5.7b.

The reader should remember that all receivers should have a post-detector filter to get rid of noise corresponding to frequencies higher than ones signal frequencies. Another specification that FM broadcasters have agreed upon is that the highest modulation frequency they will transmit is 15 kHz. Thus, one should follow the FM detector by a low-pass filter with rather sharp cut-off above 15 kHz. This will allow one to derive very significant benefit from the use of the de-emphasis filter.

EXERCISES 5.6

1. Derive the transfer function for the de-emphasis circuit and specify relations among R_1, R_2, and C, to be an effective pre-emphasizer to work in conjunction with a 75 μsec de-emphasis circuit.

ANSWERS   One needs $R_1C = 75$ µsec and $R_2 << R_1$.

2. *Derive the benefits of standard de-emphasis as follows. Assume that without the de-emphasis filter, one obtains noise power proportional to kf^2 out to 15 kHz, and thus find the noise power one would get without de-emphasis. Then, find the noise power when kf^2 is multiplied by the* power *transfer function*

$$H(f) = \frac{1}{1 + (f/f_1)^2} = \frac{f_1^2}{f_1^2 + f^2},$$

where $f_1 = 2122$ Hz. Note that the integral is "improper" in that the power of f in the numerator is no lower than that in the denominator. Hence, change it to

$$N = k \int_0^B f_1^2 \frac{f^2 + f_1^2 - f_1^2}{f^2 + f_1^2} \, df = kf_1^2 \int_0^B \left(1 - \frac{f_1^2}{f_1^2 + f^2}\right) df,$$

where B = 15 kHz and $f_1 = 2122$ Hz. Divide the result immediately above into the result without de-emphasis and evaluate the noise improvement for standard FM.

ANSWER   Ratio of noise without de-emphasis to noise after de-emphasis is

$$\text{Improvement factor} = \frac{(B/f_1)^2}{3\left[1 - (f_1/B)\tan^{-1}(B/f_1)\right]};$$

For $B = 15$ kHz and $f_1 = 2.1$ kHz, the improvement factor is 21.3, corresponding to 13.3 dB.

3. *Suppose the station engineer says, "Our listeners are getting old and losing their high frequency hearing. Let's cut off our modulation at 10 kHz." What would be the improvement factor if de-emphasis factor corner frequency is still 2.122 kHz?*

ANSWER   10.4, or 10.17 dB.

5.7   Noise in Digital Systems

5.7.1   Basic Properties of Gaussian Functions

There is a rather fortunate characteristic to the amplitude probability of noise voltages. This is illustrated in Figure 5.8, where we have graphed the probability curve for noise having an rms amplitude of 1 volt, centered at 4.0 volts.

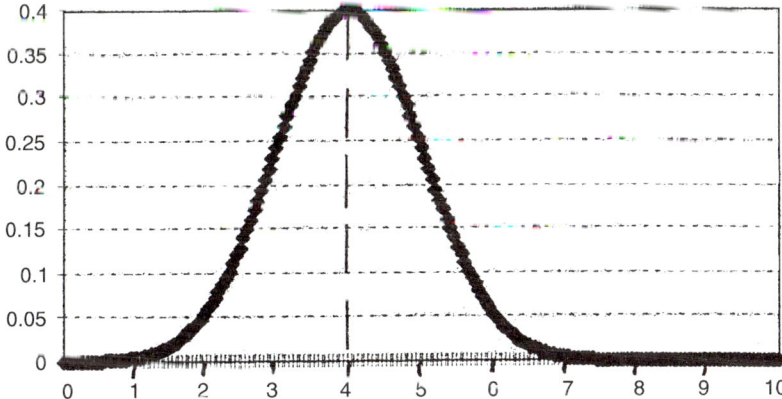

FIGURE 5.8
Gaussian probability density for mean of 4 volts, standard deviation 1 volt.

The area under the curve and between two limits of voltage corresponds to the probability that the instantaneous voltage would lie in that area. We can believe what a related calculation tells us, to wit, there is a probability of 94.5% that the instantaneous voltage would lie between 2 and 6 volts.

The nature of the "related calculation" mentioned above has to do with the area under the "Gaussian tail." One runs into difficulties if one tries to find the area using an integral in closed form. However, a numerical integral, which the author made using a not-really-long spreadsheet, gave the following numbers:

TABLE 5.1

Q(v) vs. v for rms Noise of 1v, Mean of 0

v	Q(v)	v	Q(v)	v	Q(v)	v	Q(v)
0.1	0.4602	1.1	0.1357	2.1	0.01786	3.1	9.68×10^{-4}
0.2	0.4207	1.2	0.1151	2.2	0.0139	3.2	6.87
0.3	0.3821	1.3	0.09680	2.3	0.01072	3.3	4.83
0.4	0.3446	1.4	0.08076	2.4	0.008197	3.4	3.37
0.5	0.3085	1.5	0.06681	2.5	0.006209	3.5	2.33
0.6	0.2743	1.6	0.05480	2.6	0.004461	3.6	1.59
0.7	0.2420	1.7	0.04456	2.7	0.003467	3.7	1.08
0.8	0.2119	1.8	0.03593	2.8	0.002555	3.8	7.23×10^{-5}
0.9	0.1841	1.9	0.02872	2.9	0.001866	3.9	4.81
1.0	0.1587	2.0	0.02275	3.0	0.001350	4.0	3.17

For v > 3.0, $Q(v) \approx \dfrac{1}{\sqrt{2\pi}v} e^{-v^2/2}$ is a good approximation within a few percent.

If one needs to find the inverse, a few sample values are: $10^{-5} = Q(4.265)$, $10^{-6} = Q(4.75)$, $10^{-7} = Q(5.20)$, $10^{-8} = Q(5.61)$.

Those who first worked with the Gaussian probability function also defined a function related to the area under one tail of the Gaussian function. If one would take the probability function centered at zero (zero mean), and found the total area under the curve from minus infinity up to a single limit "v," one would have something called the "error function" (erf(v)). Naturally, the total area under the curve must be unity, because if one sums the probabilities of everything that could happen, they should add to unity. Thus, if one subtracts the error function from 1.0, one obtains the "complementary error function" erfc(v) = 1 – erf(v). For some reason that is unclear to communication engineers, the definitions of error function and its complement contain what seems to be a total excess of twos. Q(v) is related to the complementary error function but with fewer twos. Let us start with some simple examples and then progress to the real communication problems.

Example 5.7.1.1

A noise voltage having Gaussian statistics has zero average voltage and an rms voltage of 1 volt. What is the probability that the instantaneous voltage has an absolute value greater than one volt?

SOLUTION   The Q-function tells the area under *one tail* of the Gaussian curve. So we look up Q(1) and read the probability that instantaneous voltage is greater than +1 volt is 0.1587. Because the Gaussian curve is symmetrical with respect to the mean value of v, there is equal probability that instantaneous voltage is more negative than –1 volt, so the answer to the problem above is $2 \times 0.1587 = 0.3174$.

EXERCISES 5.7.1.1
Work the example above for absolute values of 2 volts and 4 volts.
ANSWERS $0.0450, 6.34 \times 10^{-5}$.

Example 5.7.2.1

As a further example to illustrate application of the Q-function, let us suppose that one has a transistor biased at quiescent collector voltage of 3.0 volts, added to which is with an (amplified) noise voltage of 1.0 volt rms, $V_{cc} = 5.0$ volts, and the transistor saturates at 0.1 volt. What is the probability of the noise voltage causing cut-off and saturation of the transistor?

SOLUTION   We have added one more complication of the Q-function. If the mean voltage is not zero, we look up $Q((v - m)/\sigma)$, where m is the mean voltage (which here is the quiescent value) and σ is the rms value of the noise voltage. If the number inside the parentheses is negative, we simply look up Q of the positive number, because of the symmetry of the Gaussian curve. To find the probability of the noise cutting off the transistor, we need to evaluate the probability that the instantaneous voltage tries to go greater than 5.0 v.

That probability would be $Q((5-3)/1) = Q(2.0) = 0.2275$. The probability of noise saturating the transistor would be

$$Q((0.1-3.0)/1) = Q(-2.9) = Q(2.9) = 0.001866.$$

EXERCISES 5.7.1.2

1. *Rework the example above, changing only the noise voltage to an rms value of 0.5 volt.*

ANSWERS   Probability of cut-off is $Q(4.0) = 3.17 \times 10^{-4}$. To get probability of saturation, one needs $Q(5.8)$. If one uses the approximate formula, one gets 3.41×10^{-9}. A more accurate table of Q-functions gives $Q(5.80) = 3.316 \times 10^{-9}$, so the approximate formula is within 3.0%.

2. *Now, suppose we have a low voltage logic system in which the receiver has a system for reading out the ones and zeroes that at a decision time, if the instantaneous voltage is above 1.1 volts, reads out a logic one. If instantaneous voltage is below 1.1 volts at the decision time, a zero is read out. (This "decision threshold" would be appropriate if normal voltage for a logic zero were zero volts and normal voltage for logic one is 2.0 volts, and zeroes were slightly more probable than ones.) Let noise voltage be 0.5 volts rms. Find the probability of error for ones and zeroes here. (Hint: One uses 0 volts as the mean for logic zeroes, for the probability of errors calculation, and one needs the probability of instantaneous voltage exceeding 1.1 volts. Similarly, one uses 2.0 volts for the mean for logic ones.)*

ANSWERS P_B (probability of errors) when ones were sent is $Q(2.2) = 0.1390$.
P_B when zeroes were sent is $Q(1.8) = 0.03593$.

5.7.2   Determination of Logic Threshold Using Bayes Theorem

A mathematician named Bayes derived a theorem to optimize overall probability of error when ones and zeroes are not equally likely. A little reflection may help the reader to realize that one needs to make the more accurate decision on the higher probability digit. Symbolically, the equation for the decision threshold voltage is given by:*

$$\frac{z(v_1 - v_0)}{\sigma^2} - \frac{v_1^2 - v_0^2}{2\sigma^2} = \ln\left(\frac{P(0)}{P(1)}\right)$$

In this expression, the decision threshold is given by z, v_1 and v_0 are the normal voltage levels corresponding respectively to ones and zeroes, σ is the rms noise voltage, and P(0) and P(1) are, respectively, the probabilities of zeroes and ones. When ones and zeroes are not equally probable, the overall

---

* For derivation, see Sklar, B. *Digital Communications*, Prentice-Hall, Englewood Cliffs, NJ 1988, pp. 740–743.

probability of error is $P_B(\text{overall}) = P_B(0)P(0) + P_B(1)P(1)$, where $P_B(0)$ and $P_B(1)$ are respective probabilities of error for zeroes and ones.

Example 5.7.2

Suppose the probability of occurrence of zeroes is 0.7 and that for ones is 0.3. Logic voltage levels are, respectively, 0 and 2 volts. Noise voltage is 0.5 volts rms. Find the optimum decision threshold for minimum overall error probability and find this overall error probability.

SOLUTION   Substituting in the Bayes relation, we have

$$\frac{z(2-0)}{(0.5)^2} - \frac{2^2 - 0^2}{2(0.5)^2} = \ln\frac{0.7}{0.3} = 0.847;$$

$$8z - 8 = 0.847;$$

$$z = 8.847/8 = 1.106 \text{ volts.}$$

The threshold of 1.100 was chosen by guess, but it is seen to be very near optimum. Error probabilities now have to be determined by linear interpolation. The author gets $Q(2.212) = 0.01352$ for error probability for zeroes and $Q(1.788) = 0.0370$ for ones. Then, overall error probability is $0.7 \times 0.01352 + 0.3 \times 0.0370 = 0.0206$. Had we used the threshold of 1.0 volt, which is only optimum when ones and zeroes are equally probable, error probability would have been $Q(2.0) = 0.02275$ for ones, zeroes, and overall.

EXERCISES 5.7.2

1. In the example above, suppose that the noise voltage was 0.25 volts rms. For the same threshold of ones and zeroes, find the new optimum decision threshold.

ANSWER   1.026 volts.

2. For rms noise voltage of 0.5 volts, and for logic levels of 0 for zeroes and 2.0 for ones, find the threshold for lowest overall probability of error, if the probability of zeroes is 0.2 and probability of ones is 0.8.

ANSWERS   New threshold would be 0.799 volts. Overall probability of error is 0.0175.

5.7.3   Noise-Caused Bit Errors in Coherent Digital Systems

The analysis above would give valid results if the signal were strong enough to provide a detected signal-to-noise ratio that is greater than zero for the bandwidth and operating frequency (hence, system temperature, including receiver noise and sky temperature) than one is using. The required signal would be much greater than is obtainable in any satellite communication system, and, therefore, a detecting system which effectively increased the signal-to-noise ratio on which the receiver decides what are ones and what are zeroes, was needed. The correlating detector answered the need for what is also called a "matched filter detector" and the result might be a little startling until one comes to depend upon it. If one is feeding binary PSK into a correlation detector, the probability of error for equal ones and zeroes is*

$$P_B = Q\left(\sqrt{\frac{2E_b}{N_0}}\right)$$

where E_b is the bit energy and N_0 is noise spectral density, equal to kT_{sys}, T_{sys} is system noise temperature, including the effects of receiver noise and the antenna or "sky" noise temperature. Bit energy is given by P/R, where P is the power being received and R is the bit rate. There is an only slightly disguised factor here. Since error probability depends inversely on the square root on bit rate, if one finds inadequate error rate at a certain bit rate, one only has to demand and design for a lower bit rate. Let us first illustrate this with an example.

Example 5.7.3

We require an error probability of 10^{-5} or less in a system using binary PSK, at a system temperature of 200K, bit rate of 20,000. What power must be received?

SOLUTION   From our Q-function data, we observe that $10^{-5} = Q(4.265)$. We massage the error probability for binary PSK a bit, saying:

$(4.265)^2 = 2E_b/N_0 = 2P/RkT_{sys}$; solving this for power, we find
$P = (4.265)^2 \times 20,000 \times 1.38 \times 10^{-23} \times 200/2 = 5.02 \times 10^{-16}$ watts.

EXERCISES 5.7.3
1. In the example above, suppose one needs an error probability of 10^{-8} or better. What power is required?

ANSWER 8.69×10^{-16} watts.
2. In a binary PSK system having system temperature of 1500 K, received power of 1 femtowatt (10^{-15} watts), what bit rule can be supported for error probability of 10^{-8}?

ANSWER   3870 bits per second.
3. Now we can see that with on–off keying being fed into a correlation detector, the fact that signal is absent for zeroes simply takes the "2" out of the bit error expression. For a received power of 10^{-15}, system temperature of 1500K, and bit rate of 5000, what is the probability of error?

ANSWER 9.68×10^{-4}.

---

* Sklar, *op. cit.*, p. 166.

4. It turns out that if FSK is detected using a correlation detector, just as with OOK,
probability of error is $Q\left(\sqrt{\dfrac{E_b}{N_0}}\right)$, *assuming that the two frequencies being used for*
ones and zeroes are separated at least by the bit rate. Find the bit rate which can be
supported for system temperature of 150K, received power of 10 femtowatts, and
required error rate of 10^{-6}.

ANSWER   212 kbits sec^{-1}.

5.7.4   Noise in Noncoherently Detected FSK

The correlation detector is an example of what is called a "coherent" detector.
Now, certainly, a mildly adventurous engineer might say, "Let's just move the
FSK carrier into the FM band and see what we get as the output of the FM
detector!" There may also be other methods of noncoherent detection; let us
just bear in mind that by "noncoherent," we do *not* mean irrational or
psychotic. Sklar derives the error probability for this case as

$$P_B = 0.5\exp\left(-E_b/2N_0\right).$$

EXERCISES 5.7.4
   Work Exercise 5.7.3.4 above for noncoherently detected FSK.

ANSWER   184 kbits sec^{-1}.

6

Antennas and Antenna Systems

It is unfortunate that almost no other subject with which electronics engineers deal seems derived from so much magic and uncertainty. This writer prefers an approach which seems rational and straightforward to him, putting together building blocks of modest size to arrive at useful results. Under static conditions, the Biot–Savart law is found to be useful in predicting the magnetic fields that arise because of currents. For high frequencies, our approach here will be to use the delayed magnetic vector potential resulting from ac current. From the vector potential, it is straightforward to obtain magnetic field as the curl of the vector potential. Then, electric fields may be obtained from the magnetic fields using the differential form of Ampere's law. Knowledge of the fields surrounding an antenna may then be used to compute radiated power.

6.1   Fields from a Current Element

Figure 6.1 shows a short wire of length "d" centered at the origin and pointed along the z-axis, carrying a current for which the phasor representation might be $Ie^{j\omega t}$. Now, the essence of "delayed potential" is that the effects at some point a distance "r" away from a conductor are simply delayed by the amount of time required for waves traveling at the speed of light to make the one-way trip. Using also some consequences of Biot–Savart, the result is that the vector potential has a magnitude and time variation given by

$$\left|\vec{A}\right| = \frac{\mu Id}{4\pi r} e^{j\omega(t-r/c)}.$$

The direction of the vector $\vec{A}$ is parallel to the direction of current flow in the wire. Similar to Biot–Savart, the magnitude is inversely proportional to "r," which is the distance from wire to point where we want to compute fields. It is assumed that the medium surrounding the wire is vacuum, because waves are implied to travel at the velocity of light in vacuum, which many authors

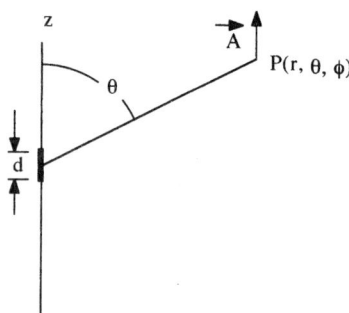

FIGURE 6.1
Vector potential from current element at origin.

call "c." It should be noted that the velocity is only attenuated in the hundredths of a percent if the antenna is in air at standard temperature and pressure. Also, note that no mention is specified of exactly what specific value of ϕ the point P is at; the reason is that a symmetry applies here, so the fields do not depend at all upon the specific value of ϕ, when the antenna element is right at the origin. However, it is necessary to be able to find the fields as a function of the spherical variable θ, so one must dissect the vector component A_z into r and θ components. Use of standard trigonometric relations yields

$$A_r = A_z \cos\theta \text{ and } A_\theta = -A_z \sin\theta.$$

Now, of course, the vector $\vec{B} = \nabla \times \vec{A}$.

However, since Ampere's law, which we use next contains instead the vector $\vec{H}$, we simply divide the μ out of the vector potential expression, and write

$$\vec{H} = \frac{Id}{4\pi r}\left(\vec{a}_r \cos\theta - \vec{a}_\theta \sin\theta\right).$$

Now, the curl expression in spherical coordinates contains in the r and θ components only derivatives of the ϕ-components, of which we have none, or partial derivatives with respect to ϕ, where there is no ϕ-variation of any components. This only leaves the ϕ-component of curl, given by

$$H_\phi = \frac{1}{r}\left(\frac{\partial(rH_\theta)}{\partial r} - \frac{\partial H_r}{\partial \theta}\right).$$

In taking the partial derivatives, we must remain aware that there is r-variation in the exponent. Hence, the result, with the $e^{j\omega t}$ time-variation understood, is

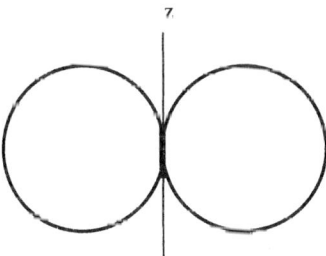

FIGURE 6.2
Vertical pattern of short dipole.

$$H_\phi = \frac{Id}{4\pi} e^{-j2\pi r/\lambda} \sin\theta \left(\frac{j2\pi}{\lambda r} + \frac{1}{r^2} \right).$$

If we now assume that our medium has zero conductivity, so that there is no conduction component of current and zero free charge and there is also no convection component, Ampere's law simply relates the curl of $\vec{H}$ to displacement current density $\varepsilon \frac{\partial E}{\partial t}$. Once one has taken the curl, one integrates with respect to t, and obtains the complete expressions for components of $\vec{E}$:

$$E_r = \frac{Id}{4\pi} \sqrt{\frac{\mu}{\varepsilon}} e^{-j2\pi r \lambda} \cos\theta \left(\frac{1}{r^2} - \frac{j\lambda}{2\pi r^3} \right)$$

$$E_\phi = \frac{Id}{4\pi} \sqrt{\frac{\mu}{\varepsilon}} e^{-j2\pi r \lambda} \sin\theta \left(\frac{j2\pi}{\lambda r} + \frac{1}{r^2} + \frac{-j\lambda}{2\pi r^3} \right).$$

All of these components may be seen to contain several powers of r; however, if one compares the various magnitudes as a function of r, one finds that for $r \gg \lambda$, the dominant component by far is the one which varies as $1/r$. Since these components dominate far from the current element, they are what mainly determine the radiation of energy by the antenna, and are called the "radiation fields." (The other fields dominate *near* the antenna hence account for the energy stored near the antenna and therefore determine stored energy and the reactance appearing in the equivalent circuit for the antenna. Determining such reactance is a very labor-intensive process, hard to do accurately, and hence will not be done here.) The form of the radiation fields makes it very straightforward to obtain the total energy radiated. The radiation fields only contain H_ϕ and E_θ; the angular variation of both of these components is $\sin\theta$. Sometimes, one plots the magnitude of a radial vector which is proportional to $\sin\theta$; this is called the "vertical* radiation pattern" for the antenna and is shown in Figure 6.2.

---

* Because radiation is independent of ϕ, the *horizontal* pattern is a circle with the antenna at the center.

We should also note that implication of the word "dipole" is that, short though it is, the wire is split in the middle and fed the opposing phases of a balanced (probably 300 ohm) transmission line.

6.2 Radiated Power and Radiation Resistance

We can obtain the radiated power by integrating the Poynting vector $\vec{E} \times \vec{H}$ over a sphere of radius r. We saw above that the fields that matter in terms of radiated power are E_θ and H_ϕ and, because the fields are perpendicular to each other, their cross-product is simply $E_\theta H_\phi$. The vectors are also in phase, so we neglect phase factors and write

$$E_\theta H_\phi = \sqrt{\frac{\mu}{\varepsilon}} \left(\frac{Id}{2\lambda r} \sin\theta \right)^2 .$$

We must remember and use that in spherical coordinates; the differential surface area is $r^2 \sin\theta d\theta d\phi$. Then, at any radius r, these r's cancel out those in the fields and our integral becomes

$$P = \int_0^{2\pi} \int_0^\pi \left(\frac{Id}{2\lambda} \right)^2 \sqrt{\frac{\mu}{\varepsilon}} \sin^3\theta d\theta d\phi .$$

The reasonably resourceful engineer should not be intimidated about integrating $\sin^3\theta$. One writes it $\sin\theta \sin^2\theta = \sin\theta (1-\cos^2\theta)$; recognizing that $-\sin\theta \cos^2\theta d\theta$ is of the form $x^2 dx$ if $x = \cos\theta$ and its integral is $x^3/3$. So, we can say that

$$\int_0^\pi \sin^3\theta d\theta = \left[-\cos\theta + \frac{\cos^3}{3} \right]_0^\pi = 1 - \frac{(-1)^3}{3} = \frac{4}{3} .$$

Next, we can say that because there is no ϕ-variation in the integral, $\int_0^{2\pi} d\phi = 2\pi$, and total radiated power is

$$\left(\frac{Id}{2\lambda} \right)^2 \sqrt{\frac{\mu}{\varepsilon}} \times \frac{4}{3} \times 2\pi .$$

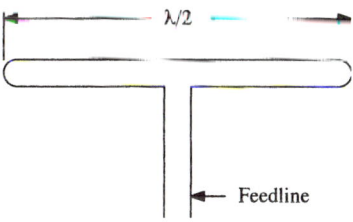

FIGURE 6.3
Folded dipole antenna.

If the dielectric is vacuum, the so-called intrinsic impedance $\sqrt{\dfrac{\mu}{\varepsilon}} \approx 120\pi$. If "I"

is the peak value of a sinusoidal current, peak power radiated is $80\left(\dfrac{I\pi d}{\lambda}\right)^2$. It

is often useful to speak of the equivalent load resistance of an antenna, to account for the average power radiated. Thus,

$$P_{ave} = I^2 R_{rad}/2,$$

where R_{rad} stands for the words "radiation resistance;" for the "short" dipole,

$$R_{rad} \approx 800\left(\dfrac{d}{\lambda}\right)^2.$$

We probably should consider a dipole to be "short" if it is equal to 0.1λ or shorter. Even at a length of 0.1λ, $R_{rad} = 8$ ohms would not be a very good impedance match for a 300 ohm line.

It has been found that such a short antenna would also have a fairly formidable reactance as part of its equivalent circuit. It works out that if one lets a dipole increase in length until it is nearly one half-wavelength in length, the humongous reactance may completely vanish and the radiation resistance approaches 75 ohms. This may have been the motivation for engineers to desire coaxial cable with 75 ohm nominal impedance, but it turns out to be wrongheaded, because the dipole needs to be fed in such a way that the symmetry plane is at ground potential, and coaxial line is most satisfactory with its outer conductor grounded, necessitating use of the transformer sometimes called a "balun" (meaning a device for connecting a balanced load to an unbalanced line. Clever, no?) One can make a structure that parallels the dipole with another wire which has no gap to connect the generator to.* See Figure 6.3.

---

* Engineers almost never get criticized for ending a sentence with a preposition. If one is Winston Churchill, it is okay to say, or roar, even "This is a type of errant pedantry up with which I will not put!"

The reader probably has seen and disrespected a folding dipole antenna if he/she has bought a new stereo tuner or receiver, as one was most likely included in the box. The lack of respect may result from the fact that the dipole may be made from the same material as the lead-in wire; to make one at home, one just cuts a length of this wire, named "twinlead," about 45% of the wavelength at the center of the FM band, solders the ends together, cuts *one* wire of the folded dipole, and connects the lead-in to those new terminals. The author once made measurements on several commercial FM antennas and compared them to a folded dipole mounted on the roof. One sparkly summer evening, the dipole was up overnight. It was oriented for best gain to the southwest, yet was found to provide decent reception from a station over 150 miles to the south*east*. The folded dipole has four times the impedance of the regular dipole; that is, the real part of 300 ohms makes an excellent match to 300 ohm twinlead. Also, since the balun transformer mentioned above has a 4:1 impedance-transforming, 300 ohm folded dipole fed by balun is a good match for 75 ohm coaxial cable.

Under some conditions (auto radios and cell phones instantly come to mind), an unbalanced antenna, called a *monopole*, may seem logical. Impedances are thus exactly one half that for the dipole, so a $\lambda/4$ monopole will have a radiation resistance around 35 ohms. Another principle we have not mentioned previously is that if the antenna shaft is somewhat "fat" in fraction of a wavelength, say 1% rather than 0.01%, the reactance will not be pronounced and will not have drastic frequency variation.

EXERCISES 6.2

1. *Find the reflection coefficients with a purely real input impedance of 35 ohms connected to a 50 ohm and to a 75 ohm coaxial line.*

ANSWERS $\Gamma_t = 0.176\angle 180°, 0.364\angle 180°$.

2. *The last time the author chopped into the lead-in for a car antenna, it looked very much like a low capacitance cable having 91 ohm characteristic impedance. Suppose 35 ohms is connected to a quarter-wavelength of 91 ohm cable. What input impedance may be expected at the other end of the cable?*

ANSWER 237 ohms.

3. *Suppose the auto antenna is telescoped to be 30 inches long. Find the wavelength at 560 kHz (KSFO in San Francisco) and calculate the radiation resistance at this frequency. Hint: Find R_{rad} for a dipole twice this long and divide by 2.*

ANSWER 3.2×10^{-3} ohms. Are you amazed *ever* to pick up an AM station?

6.3 Directive Gain

All practical antennas are directive to some extent. Oftentimes, there is expressed a "directive gain," which may be expressed in several ways. One postulates that it is possible to at least think of an "isotropic radiator," which is not really possible or even desirable to build, but that would radiate equally in all directions. So in directive gain, we say, "Consider an isotropic radiator whose Poynting vector is equal to the maximum Poynting vector of an antenna whose gain is being considered." The gain is then the ratio of the power the isotropic antenna must radiate to have the same maximum Poynting vector over the power to the antenna being considered. For simplicity, let us allow the maximum power density (Poynting vector) to be 1.0 watt meter^{-2}. Integrating this constant in the θ and ϕ directions, the isotropic antenna would require feeding it 4π watts. Now, since the fields of the short dipole vary as sin θ, the Poynting vector would be $1 \times \sin^2 \theta$. In the previous section, the resulting integral of $\sin^3 \theta$ gave a power requirement of $4\pi/3$ watts. Taking the ratio of 2π watts to $4\pi/3$ watts, we get the directive gain of the short dipole as $3/2 = 1.5$. Sometimes, gain is expressed in dB. Because the ratio is one of power,

$$dB = 10\log_{10}(1.5) = 1.76 \text{ dB}.$$

Allowing the length of a dipole to increase to a half-wavelength, one is rewarded with a pretty modest increase in gain to 1.64, or 2.15 dB. Of course, one does drastically improve impedance matching possibilities, so that one can send a lot more power out into space.

6.4 Antenna Arrays for Increased Directive Gain

A single radiating element will always have a rather limited gain. However, we can consider several antenna elements on a line or in a square in which we can control the amplitude and phase of the driving current. An arrangement in which current amplitude and phase are chosen rather deliberately and stay fixed can have much improved gain. If one has a means of controlling current phases for all elements in real time, the design is called a "phased array" and may be used to scan the skies for, for example, incoming enemy missiles.

Suppose we set out to design an array that will have a fairly high gain in a chosen direction. The general principle is to space individual radiators, which could all be short dipoles, and phase their currents in such a way that the individual elements all reinforce each other in the preferred direction. Consider Figure 6.4.

We have shown four antenna elements evenly spaced an amount "d" along the z-axis. The technical term for this arrangement is a "4-element array." Suppose we wish our beam of radiation to be strongest along the positive z-axis. Let us assume we have chosen the spacing "d" of the elements to be one quarter-wavelength. Our design scheme then will be to phase the current of each element to be 90° behind the one immediately more in the negative z-direction.

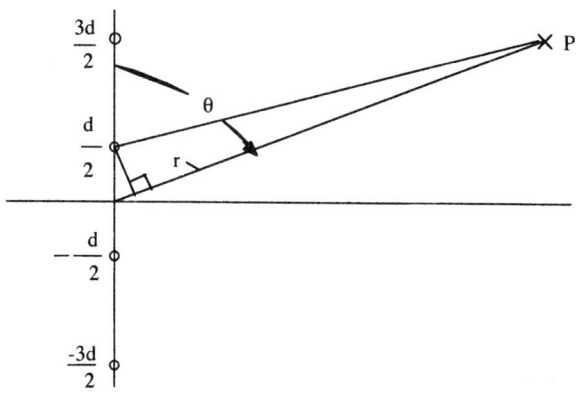

FIGURE 6.4
Antenna array with 4-elements spaced "d" from nearest others.

The scheme for delayed potential causes the effects of the individual elements to all add in phase as one moves *up* the z-axis. However, as one would move in the negative direction, the element one starts with is already a quarter period behind the next more negative one. By the time one moves the distance to the next, the effects are 180° out of phase, so there is cancellation of the effects of pairs of elements in the negative z-direction.

Having designed the array by considering the reinforcement along the z-axis, we must relate to the delayed potential mechanics. Let's say the antenna elements are all short dipoles. For a particular choice of fields, say, for example, electric fields, we can say that at some distance away each contribution will have a collection of the same constants but with individual magnitudes and phases of current and downstairs an r with a subscript saying distance from that element to the distant point P. If we number the elements from the bottom of the figure, we can write

$$E_{distant} = \text{constants} \times \left\{ \frac{I_1 \angle \theta_1}{r_1} e^{-j\pi r_1/\lambda} + \frac{I_2 \angle \theta_2}{r_2} e^{-j\pi r_2/\lambda} + \frac{I_3 \angle \theta_3}{r_3} e^{-j\pi r_3/\lambda} + \frac{I_4 \angle \theta_4}{r_4} e^{-j\pi r_4/\lambda} \right\}.$$

Next, we identify r_1, r_2, etc., in terms of the coordinates r and θ. We need to keep in mind that Figure 6.4 is a bit inaccurate in implying how close point P is; we showed it on the same page whereas if we used an accurate scale, it would be *many, many* pages away. Then, to good accuracy, we can say that

$$r_1 = r + 3d/2 \cos\theta, \; r_2 = r + d/2 \cos\theta, \; r_3 = r - d/2 \cos\theta, \text{ and } r_4 = r + 3d/2 \cos\theta.$$

But the next bit of perspective to keep is that the point P might be 10,000 km away, whereas d might be a meter or less. So the next (very good) approximation is that where r_1, etc., simply appear affecting the magnitude, in the denominator, we can simply replace the subscripted r's with the coordinate r.

The place where the subscripted values must be carefully used is in the *expo-nents* representing the phases of the contributions.

Let us recall some design choices we made above. Let us make all current amplitudes equal to 1. Also, the phase of current 1 will be the reference

$$\theta_1 = 0, \ \theta_2 = -\pi/2, \ \theta_3 = -\pi, \ \text{and} \ \theta_4 = -3\pi/2.$$

We then have four phasors and the phases below

$$\left\{ 1\angle 0 - \frac{2\pi}{\lambda}\left(r + \frac{3d\cos\theta}{2}\right) + 1\angle - \pi/2 - \frac{2\pi}{\lambda}\left(r + \frac{d\cos\theta}{2}\right) \right.$$

$$\left. + 1\angle - \pi - \frac{2\pi}{\lambda}\left(r - \frac{d\cos\theta}{2}\right) + 1\angle - 3\pi/2 - \frac{2\pi}{\lambda}\left(r + \frac{3d\cos\theta}{2}\right) \right\}.$$

If now we factor the cleverly chosen angles out of each term, we can write phases as

$$\left(-3\pi/4 - 2\pi r/\lambda\right)\left\{ \frac{3\pi}{4} - \frac{3\pi d\cos\theta}{\lambda}, \ \frac{\pi}{4} - \frac{\pi d\cos\theta}{\lambda}, \ \frac{-\pi}{4} + \frac{\pi d\cos\theta}{\lambda}, \ \frac{-3\pi}{4} + \frac{3\pi d\cos\theta}{\lambda} \right\}.$$

Looking at what we have here, we see that the first and last phases are nega-tives of each other, and likewise the second and third. We recall the identity

$$\cos x = \frac{e^{jx} + e^{-jx}}{2}.$$

We had not previously used our specification that the spacing of the antennas is d = $\lambda/4$. Since we have two pairs of such exponents, we can now write the magnitude of the E-field,

$$|E| = \text{constants} \times I/r \left\{ \cos(3\pi/4(1 - \cos\theta)) + \cos(\pi/4(1 - \cos\theta)) \right\}$$

The quantity inside the { } brackets is usually normalized and is then called the "array factor," which describes how the magnitude varies in comparison to its maximum value. We see that we have attained our design objective, which is to have its maximum at $\theta = 0$. Since cos 0 = 1, both of the cosines are unity at $\theta = 0$, so we can say the array factor is the following.

$$\frac{1}{2}\left(\cos\frac{3\pi}{4}(1 - \cos\theta) + \cos\frac{\pi}{4}(1 - \cos\theta) \right)$$

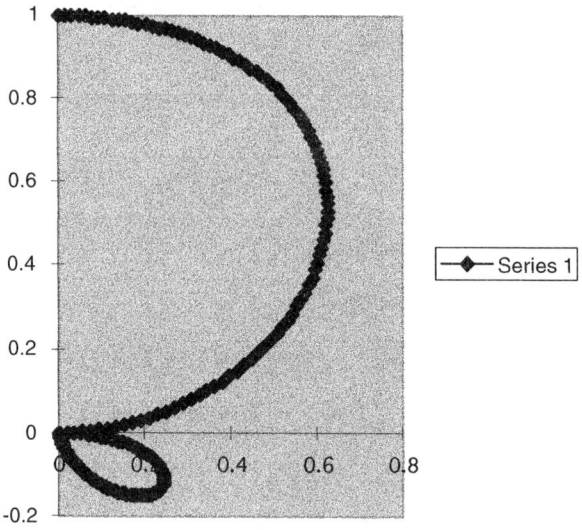

FIGURE 6.5
Radiation pattern of 4-element endfire array.

We have plotted the magnitude of the array factor in Figure 6.5, where the line shown describes this magnitude, showing simply one display of the E-field; a complete picture would join what is shown with its mirror image. There is quite a broad maximum in the θ = small degrees. Sometimes, we define a "half-power beamwidth," which describes here the rather blunt cone containing all the values of θ where the normalized E is at least 0.707. By inspection of Figure 6.5, we might estimate that the beam contained all angles out to about 60°. Actually, because we used a spreadsheet to compute Figure 6.5, we can read the exact angle as 57°, for a half-power beamwidth of 114°. We might in fact characterize this array as having uniform field intensity over quite a large portion of space. For a narrower beam, we would need an array with more elements. Suppose we consider adding just one element, keeping the distance between adjacent elements at one quarter-wavelength and the phase difference of the currents in adjacent elements at 90°. When we factor the same exponent out of each term in the array, the angular part factored out would be the exact phase of the centermost element, and there would be no term with which to pair that center term. Hence, the new array factor would have the terms

$$\text{A.F.} = k\left[1 + 2\cos\left((\pi/2)(1 - \cos\theta)\right) + 2\cos\left(\pi(1 - \cos\theta)\right)\right].$$

Again, normalizing so the array factor has a maximum value of unity, it becomes

$$\text{A.F.} = 0.2 + 0.4\left[\cos\left((\pi/2)(1 - \cos\theta)\right) + \cos\left(\pi(1 - \cos\theta)\right)\right].$$

This relation could, of course, again be plotted for all values of θ between 0 and 180° to find the new value of half-power beamwidth; however, the engineer who finds him/herself separated from his/her computer might get a quick and not-too-dirty approximation by guessing values of θ and calculating the result. Indeed, as soon as the author started this process, he decided to guess the value of 1 – cos θ, starting with 0.3, so that cosine must be calculated for 27° and 54°. The complete calculation gives 0.7915 for the result. Thus, we must try two larger angles, one of which is twice the other. Trying 30° and 60° for the angles, we get 0.7464. Trying 32° and 64°, the result is 0.714. 32.5° and 65° gives 0.7064, which is close enough. Now, we say (working in degrees)

$$32.5 = 90 \times (1 - \cos\theta), \quad \theta = 50.3°; \quad \text{beamwidth is } 100.6°.$$

EXERCISES 6.4
1. How many elements are required for an endfire antenna to have a half-power beamwidth of 75°? Keep spacing at a quarter-wavelength and driving current phases 90° apart. Remember that there is an unpaired element when the number of elements is odd, but they are all paired if the number of elements is even.

ANSWER   8-element array is down only 2.56 dB at 75° beamwidth,
   9-element is down 3.3 dB at 75°.

2. Check the performance of a 4-element endfire, with the elements one third wavelength apart and phase angles different by 120°.

ANSWERS   Array factor 0.5[cos π(1 – cos θ) + cos(0.333π(1 – cos θ)]. Half-power beamwidth is about 98°.

6.4.1   Actual Gain via Numerical Integration

Just above, we computed beamwidths because it was fairly straightforward once we had determined the array factor. To determine the actual gain, we must integrate the normalized array factor over all possible values of θ to obtain the necessary radiated power. Formally, we will have the integral of a constant in the φ-direction, which gives 2π, but the integral of sin θ times the *square* of the array factor over θ varying from zero to 180°. To do the integral numerically, let's say we will pick increments of θ one degree wide. Hence, we compute the product of the square of the array factor times sin θ at 0.5°, 1.5°, etc., multiply by a Δθ of π/180, and add all the contributions. This integral gives 2.0 for the isotropic antenna, so the gain is 2.0 divided by the sum of all contributions. The author did these manipulations for the 4-element array and got a gain of 3.6, or about 5.6 dB.

EXERCISES 6.4.1
1. Do the numerical calculations to obtain gains of the 4-element endfire array with elements one third-wavelength apart and the 8- and 9-element endfires separated a quarter-wavelength.

ANSWERS   Gain of 5.0 or 7.0 dB, gain of 8.0 or 9.0 dB, gain of 9.0 or 9.55 dB.

6.5   Broadside Arrays

In the endfire arrays, the design was such as to make all the sources interfere constructively so as to make a maximum of radiation along the line joining the center of all readapting elements. The other simple idea is to make the sources all combine constructively in a direction *perpendicular* to the line joining the centers of the elements. The word describing this effect is borrowed from naval warfare. If all of a ship's guns are firing perpendicular to the direction the ship is traveling, it is said to be firing a "broadside." Thus, the antenna array we look at next is called a "broadside array." Not too surprisingly, it is the result of feeding all the radiating elements *in phase*. Hence, we can say the currents of all elements are at the reference phase, which we can set equal to zero, and there is then no fixed angle to factor out of every term. Suppose we look at an array of eight elements, separated from each other by one quarter-wavelength. The resulting array factor then will be:

$$\text{A.F.} = 0.25\left[\cos\left((\pi\cos\theta/4)\right) + \cos\left((3\pi\cos\theta)/4\right) + \cos\left((5\pi\cos\theta)/4\right) + \cos\left((7\pi\cos\theta)/4\right)\right].$$

If we plot the magnitude of this array factor vs. the angle θ, we can get Figure 6.6.

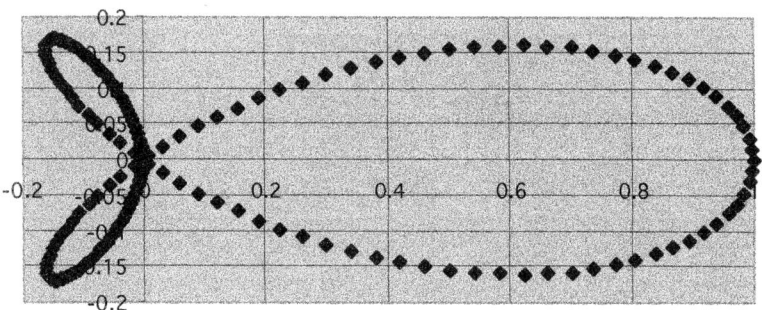

FIGURE 6.6
Pattern of 8-element broadside array.

There is a *caveat* or two, as a Latin-speaking politician might say. Perhaps we should say that the pattern represents the antenna's pickup in the direction $\phi = 0$. Except for the confusion factor it might add, perhaps we should also superimpose on this picture its mirror image reflected in the vertical axis. Because this array still has no way of varying in the ϕ-direction, the picture should also show the pattern for $\phi = 180°$. Thus, we ought not to speak of a "beam width," because the pattern of a broadside is not really a beam, but more like a Victorian lady's fan that opens up into a complete circle. Still, we can find the gain of this array by again doing a numerical integral. When this is done, the author obtains a gain of 4.16, or only about half what was obtained for the 8-element endfire. Thus, although the range of θ for which the array factor is 0.7 or greater is only about 27°, it is uniformly spread over all ϕ, so the gain is much less.

6.6 Antenna Aperture and Other Applications of Gain

We now have the tools for calculating how much power will be picked up by a receiving antenna. We may characterize a receiving antenna by its "aperture," which is the effective area on the antenna, related to the gain (an antenna's gain is the *same* whether it is used as a transmitting antenna or a receiving one. This effective area is given by

$$A_{eff} = \frac{\lambda^2}{4\pi} \times gain,$$

where the gain is given as a number (not in deciBels). We assume that each antenna is impedance-matched to its transmission line, and we can write the received power as the product of this area and the Poynting vector at the receiving antenna location. Also, we can obtain the Poynting vector at some distance from the transmitting antenna, by first calculating what it would be at that distance if the transmitted power were spread evenly over a sphere of radius equal to the distance. Then we assume that we have so aligned the antennas that the transmitting antenna has its maximum gain in the direction of the receiver, and just write the Poynting vector as

$$\wp = \frac{P_t G_t}{4\pi r^2},$$

where P_t is the power being transmitted, G_t is the gain of the transmitting antenna, and r is the distance from the transmitting antenna to the point being considered. Let us look at an example.

Example 6.6.1
A cellular phone base station has output power of 100 watts and its antenna array has gain of 9 dB. Handset antennas may be considered "short," with gain of 1.5. Assume that the operating frequency is 900 MHz. Assuming a smooth, flat earth (very unrealistic assumptions, by the way), how far away should the handset be able in order to receive 10 nanowatts?

SOLUTION We may write received power in terms of transmitted power, both antenna gains and the distance between the antennas, as

$$P_r = \frac{P_tG_t}{4\pi r^2} \times A_r = \frac{P_tG_t}{4\pi r^2} \times \frac{\lambda^2 G_r}{4\pi} = P_tG_tG_r \times \left(\frac{\lambda}{4\pi r}\right)^2.$$

One may use the form of this equation, sometimes called the "range equation" for which one has data. The reciprocal of the last term, with perhaps the wavelength written in terms of operating frequency and the velocity of light in space, is sometimes called

$$^{\text{TM}}\text{Free space loss}= \left(\frac{4\pi rf}{c}\right)^2.$$

This author has not come up with a rationale for such terminology, but the reader will not go far in space communications before he/she sees the expression and is forewarned. Substituting what we have,

$$10^{-8} = 100 \times 10^{9/10} \times 1.5 \times \left(\frac{3\times 10^8}{4\pi r \times 9 \times 10^8}\right)^2;$$

$$r = \frac{1}{12\pi} \times \sqrt{1.2 \times 10^{11}} = \frac{10^5}{\pi\sqrt{12}} = 9189 \text{ m, or about 5.5 miles.}$$

Note that conductivity of the earth, buildings, cars, and other signs of development are most likely to greatly reduce this range.

It is not really possible to count the elements on the average TV or other antenna and determine the gain; if one needs the gain, one must set up controlled environments and make radiated power measurements. There is just one structure in which fairly accurate estimates can do a good job. Consider the so-called "dish," which is basically a paraboloid (the surface generated by rotating a parabola about its axis). One must be sure that the radiating element is located at the focus of the parabola and uniformly "illuminates" the dish. Then, the effective area of the antenna is 0.55 times the area of the circular opening in the dish. If one does the calculation, one sees some truly staggering gains for large satellite dishes.

Example 6.6.2

Consider a dish having an opening of radius 1.5 m, operated at 6 GHz. Find the gain of the antenna.

SOLUTION The effective area is $0.55\pi \times (150 \text{ cm})^2 = 38{,}877 \text{ cm}^2$.
   Wavelength here is $3 \times 10^{10} \text{ cm/sec}/6 \times 10^9 \text{ Hz} = 5 \text{ cm}$.
   So, we can say $38{,}877 \text{ cm}^2 = (5)^2 \times \text{gain}/4\pi$; gain $= 38{,}877 \times 4\pi/25 = 19{,}542$.

EXERCISE 6.6

1. *Find the gain of a small dish with radius of 30 cm, if operated at 15 GHz.*

ANSWER   2827.

6.7   The Communications Link Budget

Much of the substance of the last few chapters comes together in one calculation having many factors. A transmitter, two antennas, a medium, such as space of incredible extent, and a receiver that adds some noise and decodes the message being sent, comprise a communications link. Over the extent of space, digital signaling is the only technique that stands a chance of succeeding. As was demonstrated in Chapter 5, the specification that gives the minimum criterion for success is ratio of bit energy to noise spectral density. One can state what would be ideal conditions and then one can, easily sometimes, predict what would go wrong. Early in one of the early moon trips, a video camera accidentally got pointed at the sun, with the result that we earthlings got no live TV from then on. Both antennas being used are extremely directive, with the result that very small errors in pointing may seriously degrade or completely lose the received signal. Similarly, any polarization preferences of each antenna must be rigorously adhered to, putting a considerable premium especially on the controls of the space vehicle. The power that a spacecraft can transmit depends upon its power supply; if it should happen that the spacecraft is getting its power from the sun, the farther it gets from the sun, the weaker the signal. We can generally expect that space itself is a loss-free medium; however, the distance a signal must travel through earth's atmosphere can add significant signal attenuation, especially if there is rain or snow in the signal path. Of course, atmospheric attenuation will be worse the longer the distance the signal is forced to go in the atmosphere. Straight up is of course the best for communication; if the signal path lies almost horizontally, one will experience maximum signal attenuation and will also pick up the maximum of sky noise.

In the previous section, we obtained the received signal power in terms of transmitted power, both antenna gains, and free space loss. In Chapter 5, we found we could express the necessary amount of received power in terms of a required value of E_b/N_0, which depended upon the required noise performance. We put all these items, plus a margin for error "M," into the link budget equation, which we write in terms of transmitter power required:

$$P_t = \frac{(E_b/N_0)RkT_{sys}}{G_tG_rL_0l_i}M \qquad (6.7.1)$$

To save the reader from having to search out the meanings again of all the symbols, one gets a required value for E_b/N_0 from the requirement for bit error performance and the particular from of digital modulation one is using. R is the bit rate in bits per second, and k is Boltzmann's constant. T_{sys} is system temperature, which can depend on many things, such as frequency, how far above the horizon the receiving antenna is pointed, and the receiver noise factor. The factors G_t and G_r are, respectively, gains of the transmitting and receiving antennas; L_0 is a very small number, called free space loss; L_i is a factor the author included to account for all forms of "incidental" loss due to incorrect polarization or pointing of antennas; and M is an arbitrary factor the system engineer decided upon to provide a margin for any unexpected, perhaps temporary, sources of signal loss.

Example 6.7

Calculate the transmitter power needed on the moon, about 400,000 km distant, if one requires E_b/N_0 of 20, bit rate of 20 kbits/sec, gain of 4000 for both transmitting and receiving antennas, system temperature of 100K, incidental loss factor of 0.5, a margin for error of 2.0, and operating frequency of 4 GHz.

SOLUTION   First, let us calculate free space loss:

$$L_s = \left(\frac{c}{4\pi rf}\right)^2 = \left(\frac{3\times10^8}{4\pi r \times 4\times10^8 \times 4\times10^9}\right)^2 = 2.226\times10^{-22}.$$

We substitute this into the link budget equation, Eq 6.7.1, as follows:

$$P_t = \frac{20\times2\times10^4 \times 1.38\times10^{-23} \times 100\times2.0}{(4000)^2 \times 0.5\times2.226\times10^{-22}} = 0.62 \text{ watts.}$$

EXERCISE 6.7

In the example above, keep frequency, distance, bit rate, margin for error, and incidental losses as they were, but make transmitter power = 40 watts, system temperature 250K, and required $E_b/N_0 = 10$. If the antennas have equal gain, what must it be?

ANSWER   557.

Appendix A

Calculator Alternative to Smith Chart Calculations

There once was a smart-aleck saying, "Blessed are they who go in circles, for they shall be called 'wheels'." The high frequency engineer might modify that to "for he/she is learning the Smith chart." Before the Smith chart becomes an inseparable part of the engineer's soul, it can be very confusing, and it can be very welcome to have a calculator-using alternative by which to check one's results.

A.1 Determination of Impedance from Standing Wave Measurements

Suppose we first look at determining a load impedance from standing wave measurements. While this may seem much more laborious than using one of today's network analyzers, it may be that a new engineer has joined an undercapitalized company that set up standing wave measurement equipment bought at swap meets for a couple of thousand dollars, rather than a network analyzer costing 20 to 30 times as much. Consider Figure A.1.

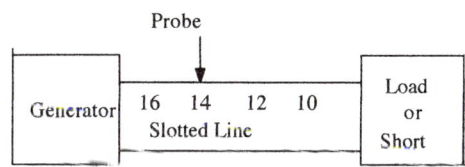

FIGURE A.1
Standing wave measurement setup.

The slotted line is a jig on which is mounted a section of coaxial cable or waveguide that has in it a longitudinal slot to permit a so-called probe to travel along it. There is a centimeter scale measuring distance traveled, which is typically numbered from right to left, if the equipment was made by the original company of Hewlett and Packard, permitting the numbers to be interpreted as distance from the load terminals. The probe picks up total wave voltage on the line, which displays standing waves. The data one records is the ratio of maximum total voltage on the line to the minimum value, which is named "standing wave ratio," sometimes given by its initials, VSWR, in which the word "voltage" was added at the beginning, and the locations of the VSWR minima. A usual first step is to put a high-quality short-circuit on the load terminals and to determine where minima are. The short guarantees there is a minimum right at the load terminals; and because standing waves repeat themselves every half wavelength, wherever there are minima on the slotted line, we know that such locations are precisely an integer number of half-wavelengths from the load terminals. This author likes to call these locations values of "d_{ref}." Thereafter, when an unknown load is connected, we can say that its impedance is precisely the input impedance presented at the reference values of "d." Our calculations must relate the impedance at the locations of minima when the unknown is connected. (The author calls these locations values of "d_{min}" to the value at d_{ref}.) Theory tells us that the impedance at a standing wave minimum is purely real and given by characteristic impedance of the line divided by the standing wave ratio. The equation giving us the numerical calculation of load impedance is

$$Z(d) = Z_0 \frac{1 - |\Gamma| e^{j4\pi(d_{ref} - d_{min})/\lambda}}{1 + |\Gamma| e^{j4\pi(d_{ref} - d_{min})/\lambda}}.$$

Example A.1

Let us do an example illustrating use of this formula. Suppose, with the short-circuit on the load terminals, one found minima (values of d_{ref}) at 10, 12, and 14 cm. Then, with the unknown load connected, one measured a VSWR of 2.5, with minima 9.6, 11.6, and 13.6. (The envious beginner will note the results of flawless technique, with well-behaved, precisely spaced data points.)

SOLUTION   Now, Γ is the reflection coefficient of the unknown load, and is given in terms of the standing wave ratio as

$$|\Gamma| = \frac{VSWR - 1}{VSWR + 1};$$

hence, in this example, we get

$$|\Gamma| = (2.5 - 1)/(2.5 + 1) = 3/7.$$

The next thing we note is that the distance between adjacent minima on the slotted line is one half-wavelength, or $\lambda/2$. Hence, here

$$\lambda = 2(12 - 10) \text{ cm} = 4 \text{ cm.}$$

Now, we can figure that imaginary exponent of e as

$$4\pi(10 - 9.6) \text{ cm}/4 \text{ cm} = 0.4\pi \text{ radians.}$$

If one prefers, he/she can think of the exponent as 72°. Hence, one enters one's calculator (which is able to deal glibly with complex numbers) with a complex number having magnitude of $3/7 = 0.4286$ and an angle of 0.4π radians or 72°, depending upon personal preference, subtracts it from $1\angle 0°$, also *adds* it to $1\angle 0°$, divides the former result by the latter, and obtains the normalized load impedance as

$$\frac{Z_L}{Z_0} = 0.7964\angle -0.7847 \text{ (radians)} = 0.5635 - j0.5628.$$

EXERCISES A.1

1. *When a new unknown impedance is connected to the slotted line above, one observes a VSWR of 3.0, with minima at 10.5, 12.5, and 14.5 cm. What is the unknown normalized load impedance?*

ANSWER $0.6 + j0.8$.

2. *When yet another unknown load is connected to the slotted line, one measures a VSWR of 2.0, with minima at 9.5, 11.5, and 13.5 cm. Find the load impedance.*

ANSWER $0.8 - j0.6$.

A.2 Series or Shunt Reactance Location and Value to Match Impedance

The calculations that really make the young engineer's head spin seem to be the impedance-matching calculations. Let us first consider the case where we plan to insert reactance in *series* with the transmission line. This geometry would be most obvious in microstrip lines. In principle, it is straightforward to expect to etch a gap at the appropriate location in the signal-bearing conductor. Then, a surface-mount inductor or capacitor can be soldered across the gap, which should then match the line if our previous calculations have been correct.

What makes it possible to determine where the series reactance must be placed is the circumstance that, assuming there is negligible attenuation in the amount of line over which matching is done, the *magnitude* of input reflection coefficient stays fixed at its value at the load, so one rotates on the chart

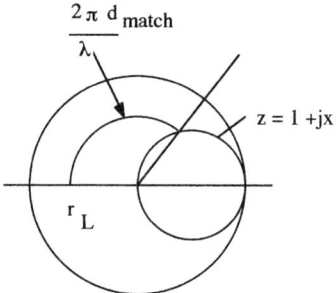

FIGURE A.2
Rotation from load resistance to matching point.

in a circle of constant radius. Phase, however, decreases proportionally to the distance one moves along the line to get to the match point. The phase retardation of the reflection coefficient is exactly *twice* the amount of phase shift that a signal in either direction would undergo. Thus, if one moves 0.1λ along the line, phase of the reflection will be retarded $0.2 \times 360°$, or $72°$. Some of the geometry involved appears in Figure A.2. Repeating what we assumed for emphasis, we said we will put the matching reactance in *series* with the line. Hence, the point at which we can match the line is where normalized input impedance will be $1 \pm jx$. Hence, adding $\mp jx$ in series with the line eliminates the total reactance, giving a purely real input impedance equal to the characteristic impedance of the line.

Because we know that *magnitude* of reflection coefficient did not change as we moved along the line to the matching point, we can say that

$$\left|\Gamma_L\right| = \left|\frac{1+jx-1}{1+jx+1}\right| = \left|\frac{jx}{2+jx}\right| = \frac{x}{\sqrt{4+x^2}}.$$

Solving for $|x|$, we get

$$|x| = \frac{2\left|\Gamma_L\right|}{\sqrt{1-\left|\Gamma_L\right|^2}}.$$

If it should happen to be our plan to match impedance by going to the location where we can put a reactive *admittance* from signal-bearing conductor to ground, we need to move from the load terminals to where the normalized admittance is equal to $1 \pm jb$. In terms of normalized admittance, we can write the reflection coefficient at the matching point as

$$\Gamma_{match} = \frac{1-(1+jb)}{1+1+jb}.$$

Solving for the value of b, we get the *same result as for x!!!*

$$|b| = \frac{2|\Gamma_L|}{\sqrt{1-|\Gamma_L|^2}}$$

However, the reader should be *scrupulously careful* to shift phase the minimum amount that brings one to the correct matching point. Let us do one example in which we use series reactance and one in which we add shunt susceptance. (The reader is reminded that susceptance is the word for the reciprocal of reactance.)

Example A.2

A 20 ohm real load is to be matched to a 50 ohm line at a frequency of 1 GHz. How far must we move from the load terminals, and what is the type and magnitude of reactive circuit element which should be added?

SOLUTION   We start by figuring $\Gamma_L = \dfrac{20-50}{20+50} = \dfrac{-30}{70} = 0.4286\angle 180°$.

We find that Figure A.2 closely corresponds to where we enter the Smith chart and where and how far we move. Our handy formula says that we will intersect the $1 + jx$ circle at

$$X = \frac{2\times(3/7)}{\sqrt{1-(3/7)^2}} = \frac{6/7}{\sqrt{\dfrac{49-9}{49}}} = \frac{6}{\sqrt{40}} = 0.9487.$$

Note that this was a normalized input reactance and, being above the axis of the Smith chart, is positive. Hence, we must add –j0.9487 to match the line. Hence, we need a capacitor, given by

$$0.9487 = \frac{1}{\omega C Z_0}; \quad C = \frac{1}{2\pi \times 10^9 \times 0.9487 \times 50} = 3.31 \text{ pF}.$$

This is not, of course, a standard RETMA value for a capacitor, but 3.3 pf would be, so that if one needs, he/she could run one's entire inventory through a network analyzer until satisfied. If the exact capacitor were used, the generalized reflection coefficient at the matching point will be

$$J' = \frac{0.9487\angle 90°}{2+j0.9487} = \frac{0.9487\angle 90°}{2.214\angle 25.38°} = 0.4286\angle 64.62°.$$

At the load, the reflection coefficient phase was 180°, but at the match point we need to be down to 64.62°, a decrease of 180 − 64.62 = 115.38°. This leads to what we needed to move from the load an amount,

$$d_{match} = (115.38/720)\lambda = 0.160\ \lambda.$$

Example A.3

Suppose we are forced to match with a *shunt* reactive element. (This could be forced upon one if one is working in coaxial cable or waveguide.) To see what is happening, it might be helpful to look at Figure A.3, which is simply Figure A.2 with the circle y = 1 ± jb (drawn dashed).

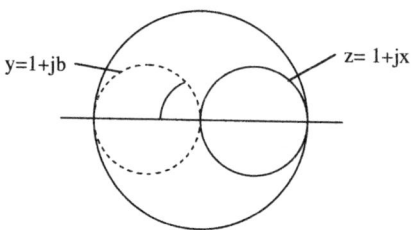

FIGURE A.3
Smith chart showing destination circles for impedance-matching using series and shunt reactances.

SOLUTION   We first intersect the y =1 + jb circle in the upper half of the chart, where reactance is positive, but *susceptance is negative*. Thus, our first chance to match impedance using shunt susceptance is found where

$$y_{in} = 1 - j0.9487.$$

Generalized reflection coefficient at the point of the match is written in terms of normalized admittance as

$$\Gamma_{match} = \frac{1-y}{1+y} = \frac{1-(1-j9487)}{2-j0.9487} = \frac{0.9487\angle 90°}{2.214\angle -25.38°} = 0.4286\angle 115.38°.$$

Now we can get the distance from the load to the matching point by subtracting this angle from 180°. We get

$$d_{match} = (180 - 115.38)°/720° = 0.8975\ \lambda.$$

At the matching point, the normalized input admittance had a negative imaginary part, so impedance matching requires a positive admittance, or a capacitor again. We have

$$0.9487 = \frac{\omega C}{Y_0} = \omega C Z_0; \quad C = \frac{0.9487}{2\pi \times 10^9 \times 50} = 3.02 \text{ pF}.$$

EXERCISES A.2

1. It is desired to match an 80 ohm load to a 50 ohm line. Find the distance from the load to put series and shunt reactances, and what circuit elements, of which value, are needed.

ANSWER At 0.1065λ from the load, put a 3.77 nH inductor in series with the line; or at 0.1435λ from the load, connect 6.77 nH from the line to ground.

2. Still in 50 ohm line, we have a load reflection coefficient of $0.3333\angle-90°$. How far from the load does one put series and shunt reactances to match the line?

ANSWER Put a series capacitor of 4.50 pF 0.277λ from the load or a shunt capacitor of 2.25 pF a distance of 2.23λ from the load.

A.3 Short-Circuited "Stubs" for Impedance-Matching

In the dark but technologically exciting days just before, during, and after World War II, chip inductors or capacitors were not available, and if one needed to match a coaxial line, about the only way the engineer had of getting a specified reactance was to short- or open-circuit a carefully chosen length of transmission line. Borrowing nomenclature from the enthusiastic radio amateurs who learned a lot by doing, these shorted or open lengths of transmission line were called "stubs." Actually, the required length of a stub can be obtained with great ease from the Smith chart, once one swallows ones fears. However, we will provide a formula and some useful precautions.

It may be more common that shorted stubs were used rather than open ones; the reason is that a wire hanging out in space does its best to become an efficient antenna, and thus its radiation resistance may move the impedance significantly away from purely imaginary to a complex one. For the reasons stated, we shall speak only of shorted stubs. These also will always be *shunt* reactances. The normalized admittance of a shorted stub is given by

$$y = Y_{in}/Y_0 = -j/\tan(2\pi l_{stub}/\lambda).$$

Now, there is a pitfall here with all calculators the author knows and many computer programs. If one needs to invert $\tan^{-1}(-$ number$)$, most calculators give an angle in the fourth quadrant. What one must have is the value in the second quadrant. What one does is simply add $180°$ or π radians to the erroneous and wrong-headed answer provided by the calculator. Let's just try a couple of examples.

Examples A.3

Find the length of a shorted stub needed to provide y = –j1 and +j1.

SOLUTIONS   First, we have

$$-j = -j/\tan\left(2\pi l_{stub}\right)/\lambda\,;$$

$$\tan\left(2\pi l_{stub}\right)/\lambda = 1;$$

$$\left(2\pi l_{stub}\right)/\lambda = \pi/4\,;$$

$$l_{stub}/\lambda = \pi/8\pi\,;$$

$$l_{stub} = \lambda/8.$$

Second, we have

$$-j = -j/\tan\left(2\pi l_{stub}\right)/\lambda\,;$$

$$\tan\left(2\pi l_{stub}\right)/\lambda = -1;$$

$$\left(2\pi l_{stub}\right)/\lambda = -\pi/4.$$

Now, of course, taken literally this answer would specify a stub of negative length, which is of mysterious practical value. So, we add π to our answer $-\pi/8$, obtaining $3\pi/8$, so the length of the stub must be $3\lambda/8$.

EXERCISES A.3

1. Find the length of shorted stubs to provide b = –5, –2, –0.5, 0.5, 2, and 5.

ANSWERS   0.0316 λ, 0.0738 λ, 0.1762 λ, 0.3238 λ, 0.4262 λ, 0.4684 λ.

Index

A

Amateur radio enthusiasts, method of choice for, 88
AM broadcast signals, 92
Ampere's law, 7, 136
Amplified noise figure, 112
Amplifier(s)
  fixed-tuned, 66
  IF, 71
  noise factor of cascaded, 113
  power gain of, 116
  requirements, RF, 68
  RF, 69, 71
Amplitude modulation, as double sideband with carrier, 90
Analog modulation, 103
Analog multipliers, 84
Angle modulation, 82
Antenna(s)
  aperture, 147
  array, 141, 142
  gains, 149
  handset, 147
  pointing of, 150
  polarization preferences of, 149
Antennas and antenna systems, 135–150
  antenna aperture and other applications of gain, 147–149
  antenna arrays for increased directive gain, 141–146
  broadside arrays, 146–147
  communications link budget, 149–150
  directive gain, 140–141
  fields from current elements, 135–138
  radiated power and radiation resistance, 138–140
Armstrong modulator, 99
Array(s)
  antenna, 143
  broadside, 146

endfire, 144
factor, 143, 144, 146
Attenuation
  constant, 12, 13
  neglecting, 19
Available noise power, 111

B

Balun, 7, 15, 139, 140
Baseband signal, 10
Base spreading resistor, 26
Bayes theorem, 131, 132
Beam width, 147
Biot–Savart law, 135
Bipolar transistors, high frequency models of, 25
Bit errors, noise-caused, 132
Boltzmann's constant, 109, 150
Broadside arrays, 146

C

Capacitance, 5, 9
Capacitive iris, 56, 57
Capacitive reactance, 2
Capacitor
  official, 2
  reactance of parallel, 72
  series, 43
  surface mount, 37
  voltage, 78
Carrier
  amplitudes, constellation showing, 107
  frequency, 82, 84, 98
  power, 123
  signal, 83, 86
Cascaded amplifiers, noise factor of, 113
Ceramic filters, 64
Circuit(s)
  for coaxial line, 11

159